The Silence of Fallout

The Silence of Fallout:
Nuclear Criticism in a Post-Cold War World

Edited by

Michael Blouin, Morgan Shipley and Jack Taylor

The Silence of Fallout: Nuclear Criticism in a Post-Cold War World,
Edited by Michael Blouin, Morgan Shipley and Jack Taylor

This book first published 2013

Cambridge Scholars Publishing

12 Back Chapman Street, Newcastle upon Tyne, NE6 2XX, UK

British Library Cataloguing in Publication Data
A catalogue record for this book is available from the British Library

Copyright © 2013 by Michael Blouin, Morgan Shipley and Jack Taylor and contributors

All rights for this book reserved. No part of this book may be reproduced, stored in a retrieval system, or transmitted, in any form or by any means, electronic, mechanical, photocopying, recording or otherwise, without the prior permission of the copyright owner.

ISBN (10): 1-4438-4479-9, ISBN (13): 978-1-4438-4479-6

In memory of Paul Boyer

TABLE OF CONTENTS

List of Illustrations ix

Preface x
John Canaday

Introduction 1
The Silence of Fallout
Michael J. Blouin, Morgan Shipley and Jack Taylor

Chapter One 16
"What Works": Instrumentalism, Ideology, and Nostalgia in Post-Cold War Culture
Jeff Smith

Chapter Two 45
Specters of Totality: The Afterlife of the Nuclear Age
Aaron Rosenberg

Chapter Three 59
Queer Temporalities of the Nuclear Condition
Paul K. Saint-Amour

Chapter Four 81
Apocalypse Networks: Representing the Nuclear Archive
Bradley J. Fest

Chapter Five 104
Cut to Black: Nuclear Criticism in a Post-September 11th America
Joseph Dewey

Chapter Six 122
The Pixilated Apocalypse: Video Games and Nuclear Fears, 1980-2012
William Knoblauch

Chapter Seven 143
Depictions of Destruction: Post-Cold War Literary Representations
of Storytelling and Survival in the Nuclear Era
Julie Williams

Chapter Eight 166
Allegories of Hiroshima: Toward a Rhetoric of Nuclear Modernism
Mark Pedretti

Chapter Nine 192
War as Peace: Afterlives of Nuclear War in David Foster Wallace's
Infinite Jest
Jessica Hurley

Chapter Ten 211
The Hunger Games: Darwinism and Nuclear Apocalypse Narrative
in the Post-9/11 World
Patrick B. Sharp

Chapter Eleven 230
Legacy Waste: Nuclear Culture After the Cold War
Daniel Cordle

Chapter Twelve 250
In a dark wud: Metaphors, Narratives, and Nuclear Weapons
John Canaday

Contributors 272

Index 276

LIST OF ILLUSTRATIONS

Fig. 6-1: Screen capture from the 1983 film *Wargames*, set in NORAD.
Fig. 6-2: Screen capture from *Nuclear War*, which pokes fun at real world political leaders.
Fig. 6-3: "If you see the flash, duck and cover!" "Vault Boy" from *Fallout*.
Fig. 6-4: Screen capture of detonating a nuclear warhead in *Fallout 3*.
Fig. 6-5: Screen capture from Introversion Software's *DEFCON* (2006).

PREFACE

JOHN CANADAY

For more than a decade, from the opening of Site Y of the Manhattan Project in 1943 to the revocation of his security clearance in 1954, J. Robert Oppenheimer probably knew more about nuclear weapons—and nuclear issues generally—than anyone. As director of the Los Alamos Laboratory, he oversaw every aspect of the first bombs' design and construction; indeed he was often credited with having more insight into the details of the work being done in the project's many subdivisions than the individuals who were doing the actual work. After the war, as Chairman of the General Advisory Committee of the Atomic Energy Commission, he was in touch with everyone who was anyone in the (western) world of nuclear issues, including scientists, politicians, and military leaders, and he had access to classified details of the technical, military, and political aspects of the development and deployment of nuclear weapons by the United States.

In 1954, despite the services he had rendered to the United States, his enemies—and his outspokenness (sometimes acerbic, always brilliant) had inspired a number of personal and ideological grudges—successfully engineered a kangaroo trial before a panel of three judges. Over the course of the four week "hearing," Oppenheimer's previous association with communists and irregularities in his exchanges with security personnel during the war years were used as a pretext for revoking his security clearance just weeks before it would have expired anyway.

Though the panel's two to one decision may have been, as Ward Evans, Chair of Northwestern University's chemistry department, said in his dissenting opinion, a "black mark on the escutcheon of our country,"[1] Oppenheimer seemed to feel it, with some justification, even more strongly as besmirching of his character. He continued to do worthwhile, even important work—writing, lecturing, and directing the Institute for Advanced Study—and he avoided expressions of anger or betrayal; but behind the public façade, he nursed a profound bitterness—to the extent that Isidor Rabi, his friend and colleague, said, "I think to a certain extent it actually killed him, spiritually."[2]

Broken or not, after losing his security clearance, Oppenheimer became more reticent in his comments on nuclear issues. A few years later, for instance, when a network interviewer asked, "Could you tell us what your thoughts are about what our atomic policy should be?" the man who had once been among the most qualified individuals to speak on these matters answered, "No, I can't do that. I'm not close enough to the facts and I'm not close enough to the thoughts of those who are worrying about it."[3] His tone was muted, his voice weary and tinged, perhaps, with sadness, yet also firm, almost defiant, as though he was determined not to let anyone—including himself—forget that the United States had deigned to find him unworthy.

It is possible that Oppenheimer did not really believe his own disclaimer. He may have allowed bitterness or discouragement at the state of U.S. nuclear policy—by then thoroughly committed to a strategy of Mutually Assured Destruction based on the proliferation of fusion bombs, whose development he had opposed—to dictate his retiring response. But as fellow outsiders, excluded from the approved and informed coterie of nuclear policy makers, humanist scholars can hardly afford to ignore Oppenheimer's assertion. If even the "Father of the Atomic Bomb" believed he could not comment usefully on nuclear weapons without highly specialized, classified knowledge, how can scholars in the humanities, with no access to such information—lacking security clearances or even, in general, any scientific background—hope to add anything useful to our collective understanding of nuclear issues? Wouldn't "Nuclear Studies" conducted by such armchair quarterbacks be no more than a dilettantish and presumptuous hobby?

These questions should give humanists[4] pause. But before going so far as to imitate Oppenheimer's reticence, it is worth using that pause to consider seriously what the humanities might really contribute to the study of nuclear issues. Part of the challenge here is that "the humanities" denotes a rattle bag of disciplines, often poorly defined and always fragmented into competing, sometimes incompatible sub-specializations. Any working definition needs to acknowledge these tensions and contradictions—indeed incorporate them as an essential feature—while catching something of that elusive essence that the term recognizes as shared by each of the disciplines it embraces. Though no definition will be entirely satisfactory, I believe we can make useful headway by proceeding on the understanding that the (scholarly) humanities consist of those analytical (as opposed to empirical) activities that strive to construct satisfying representations of human experience and its social and cultural expressions.

The weakest link here, of course, is the word "satisfying," but it is worth embracing. The conditions that count as "satisfying" are legion, which makes for awkward systematizing but reflects the humanities' essential diversity of aims and uses. In general, people seem to be satisfied by the "completion" of patterns—sequences of linked elements that imply additional elements. I place "completion" in quotes because our satisfaction may derive from experiencing a fulfillment, a surprising redirection, or sometimes a purposeful frustration of the expectations raised by the original pattern. Music, rhymes, crossword puzzles, jokes, math problems, baseball games, explanations, sex. Lacking experimental verification of their claims, humanist analyses depend on appeals to our apparent need for closure, including, with real or asserted prominence, that of logical structures of thought.

The words "representations" and "expressions" are also loose and insufficiently distinct from one another. Again usefully so: humanist work is concerned with a wide variety of symbolic activities, ranging from the representational to the abstract, and it is often, implicitly or explicitly, self-referential, sometimes to the point of being more concerned with its media than either their content or context. This can be part of the usefulness of humanist work; it can also be a major weakness: at best because its results are not reproducible, depending on style more than substance; at worst because it invites sloppy thinking. While the sciences seek to produce symbolic representations of physical phenomena based on objective, repeatable observation and manipulation, the humanities concentrate on phenomena that are already symbolic, exploring representations by means of subjective, intuitive observations. While scientists strive to construct representations of the concrete, phenomenal world that are as abstract as possible—ideally encoded in mathematical language—in order to minimize traces of their own subjective agency, humanist representations inhabit that subjective middle ground, managing neither strict fidelity to the phenomenal world nor a full inductive abstraction of it, but instead remaining enmeshed in the symbolic media of their objects of study. While the sciences seek to transcend the quirks of individual human experience, the humanities celebrate, indeed depend upon, their imbrication in these vagaries. Including both "representations" and "expressions" in our description of the humanities emphasizes these entanglements in the (as yet) unquantifiable realm of feelings and intuitions.

In the context of this brief description of the humanities, we can now turn to consider a pair of questions: What is this "nuclear" thing humanists want to study? and, Why are they drawn to it if it lies outside their disciplinary purview(s)? It may seem self-evident that "Nuclear Studies" center on the examination of technologies powered by nuclear processes, and while this is broadly true, it oversimplifies the subject in ways that imply scientific knowledge as a prerequisite. In practice, "Nuclear Studies" may focus on a wide range of issues, including theoretical, experimental, technological, historical, military, political, social, and cultural, singly or in conjunction, such as fission cross-sections and neutron energy spectra, the development of cyclotrons and betatrons, the innovations of water boiler reactors and explosive lenses, the birth of big science, the evolution of the military industrial complex, the history of the SALT treaty talks, the spread of Mutually Assured Destruction as a strategic principle, the representation of nuclear war in literary and artistic productions, and the use of nuclear imagery to market everything from candy to cars.

In order to avoid losing sight of the trees in the forest, it is useful to distinguish the two poles marking the terrain "Nuclear Studies" seeks to analyze: A) the technological artifacts themselves—particularly weapons—that derive energy from the fission or fusion of atomic nuclei, and B) the inscription of these artifacts in social and cultural contexts. This distinction can help us clear up the confusion regarding the role of the humanities in "Nuclear Studies." On the one hand, it acknowledges the limitations of a humanist approach: without specialized, technical knowledge, no-one can hope to address A. On the other hand, humanists have a home field advantage when it comes to B. Failure to distinguish between these poles leads to a confused state of affairs in which humanists assume they are not qualified to undertake "Nuclear Studies" and scientists assume they are. In fact, it is just as reasonable to ask whether scientists are equipped to address the social and cultural inscriptions of nuclear artifacts as it is to question whether humanists are competent to examine their technological aspects.

Perhaps it is not surprising that many nuclear scientists have acted as though their specialized knowledge of A has automatically qualified them to speak to B. Subatomic physics is an abstruse topic, involving a rarified vocabulary and mathematical facility, whereas social and cultural contexts would seem to be domains of common sense. But why do politicians and military leaders (in general), though lacking scientists' technological credentials, seem to share their willingness "to boldly go"? Masters of neither scientific nor scriptive pole, soldiers and statesmen occupy the middle ground, tasked with "making things happen." Rightly or wrongly, to our weal or woe, it is the nature of their occupations that they must make decisions and push for action. We might hope that they would avail themselves of experts on either side, and they have been known at least to gesture in deference to scientific knowledge; but they have shown little or no inclination to consult with humanists. Perhaps they sense our hesitance.

Curiously, humanists feel no similar diffidence when approaching other subjects that also have a clear technological basis, such as film studies, even though I think it's safe to say most of us don't understand the function of silver halide in film stock, the optics of an IMAX projector, or the neurological processes that translate a series of still images into the illusion of a moving picture. This lack of self-doubt reflects both a feeling that film technology is less abstruse than its nuclear cousin and the fact that movies are more obviously social and cultural entities—and easily available ones at that—whereas nuclear weapons exist as immensely sophisticated technological artifacts accessible only through various degrees of specialized separation. Even as symbols—the way in which most of us know them—nuclear weapons are almost entirely absent, experienced only as theoretical signifiers pointed to by other symbols.

The general absence of nuclear weapons (and, to a lesser extent, other nuclear artifacts) from our field of experience is perhaps the crucial issue. Film studies is characterized by the apparent presence of an object of study. We can go to a movie theater, buy a DVD, download a film from iTunes. But we can't see, touch, or buy a nuclear weapon, and most of us will never visit a nuclear power plant or operate a nuclear radiograph. It is this absence of an object in "Nuclear Studies" that makes the discipline so fraught for humanists: how do you study what you can't see, touch, taste, smell, or hear? Our practical distance from nuclear artifacts mirrors and intensifies our sense of epistemological distance.

It is important to recognize that this absence—though in different forms—also characterizes the relationship between nuclear scientists and their objects of study. Whole atoms cannot be grasped by the senses, much less nuclei, and physicists have had to become masters of indirection, inference, and metaphor (in both verbal and mathematical forms). Ironically, this distance led many scientists involved in the initial development of nuclear weapons to turn to literary sources and devices in their efforts to represent, even to themselves, the things they were working on. Thus there is an essential commonality of experience in scientific and humanistic involvement in "Nuclear Studies." Our authority, whether as humanist or scientist, derives not from privileged proximity, but from an ability to read with subtlety and sensitivity the symbolic inscriptions through which nuclear phenomena are written in natural and social contexts.

Humanists, therefore, are not necessarily interlopers in a realm reserved for the technological cognoscenti. As long as they focus their efforts in the vicinity of the pole that marks their area of training, humanist expertise in the identification and interpretation of social and cultural representations can play a crucial role in forming a better understanding of nuclear technologies—which exist outside the lab or bunker precisely in the form of such representations. Though some individuals will be able to venture further from their native territories, scientists as well as humanists would do well to remain mindful of their limitations.

Of course expertise is itself a vexed question, with no clear rules for determining who has it and who doesn't. Academic degrees, student evaluations, and peer-reviewed publications form a rough set of criteria, but each time an individual ventures to learn something new, she is crossing back over the (imaginary) boundary between expert and student. This dynamic becomes particularly significant when an expert in one field ventures into territory that lies between disciplines or that, even more dangerously, might be claimed by another discipline altogether. Sokal's Hoax and the ensuing Science Wars illustrate some of the pitfalls and offer useful cautions to specialists from either pole. When venturing into new territory, one would be wise to respect the authority of the locals and, in fact, to seek out their guidance and collaboration. In these circumstances, peer review becomes even more essential, as does taking seriously our responsibility to learn the work of those who have preceded us and to recognize our own limitations.

Oppenheimer understood this better, perhaps, than anyone. It was his extraordinary breadth of knowledge, which included a deep grounding in the humanities—a mastery of French literature, of Metaphysical poetry, of sacred Sanskrit texts, of Indonesian cooking—in addition to his expertise in physics, that allowed him to speak wisely and stylishly to both poles of "Nuclear Studies." Even Oppenheimer, however, retained a sense of his own limitations, and when his security clearance was revoked and he lost access to what he believed to be essential technical knowledge, he curtailed the scope of his comments. During the heady days of his ascendancy, when he was one of the most influential analysts of nuclear issues, he did not shy away from sharing his opinions. Yet even then his pronouncements were models of rhetorical grace and subtlety. Even then he sensed that his understanding of nuclear issues was as much—or more—a matter of style than substance.

Describing the difficulties involved in finding a workable political solution to the challenges involved in international nuclear control, for instance, Oppenheimer asserted in 1948:

> The problem of doing justice to the implicit, the imponderable, and the unknown is of course not unique to politics. It is always with us in science, it is with us in the most trivial of personal affairs, and it is one of the great problems of writing and of all forms of art. The means by which it is

> solved is sometimes called style. It is style which complements affirmation with limitation and with humility; it is style which makes it possible to act effectively, but not absolutely; it is style which, in the domain of foreign policy, enables us to find a harmony between the pursuit of ends essential to us, and the regard for the views, the sensibilities, the aspirations of those to whom the problem may appear in another light; it is style which is the deference that action pays to uncertainty; it is above all style though which power defers to reason.[5]

The force of "the implicit, the imponderable, the unknown" in science and in human affairs can hardly be illustrated more clearly than by nuclear power. It is ironic that Oppenheimer would forget this after having his security clearance revoked—forget there is more to say about nuclear issues than can be included in a security briefing, forget the importance of style and his own sensitivity to it. But perhaps his later reluctance to assert himself was less forgetfulness than a loss of faith in the force of his own personal style, which had so notably failed to make the powers that put him on trial defer to reason.

Of course Oppenheimer was (and is) not the only person to forget the two poles of "Nuclear Studies." Style is precisely what politicians and military strategists discount. Pressed by their need to devise literal applications for nuclear weapons, for instance, they fall into the trap of behaving as if these weapons are accessible to the expert, are amenable to reason, are objectively knowable. In doing so, they ignore the way nuclear weapons function in society and the buttons in us they can push, focusing instead solely on their ability to destroy an enemy and the attendant questions of manufacture, maintenance, situation, security, and delivery. Even if nuclear weapons existed only as technological artifacts, this approach would be imperfect; but as cultural entities, they defy such reductive, statistical manipulation.

If we are to escape the hazards of purely literal readings of nuclear technologies, we need to recognize the importance of style. We must attend to the ways in which we represent these technologies, not just to the ostensible content of those representations—for the *ways* we say things also encode meaning, whether we are aware of it or not. We must remember, with Oppenheimer, that meaning-making is not only a denotative process but a connotative one as well. Our understanding of nuclear weapons must go beyond megatons and spark gap switches, our grasp of nuclear power plants reach past megawatts and zirconium cladding, because these technologies are not inevitable, universal manifestations of natural laws but socially specific human manipulations of forces we understand only imperfectly, the consequences of which lie far beyond our ability to predict.

In short, our actions should pay deference to our uncertainties, not hide them in the name of security under a spackle of secrecy, no matter how expertly applied. In expressing deference, we would be following the example of physicists in the 1920s who acknowledged the fundamental uncertainties involved in their efforts to understand the subatomic realm. Indeed for them this was only the beginning, and they went further, recognizing in Werner Heisenberg's Uncertainty Principle the essential wisdom inherent in an acceptance of the limitations of objectivity. It is time for us, likewise, to embrace the metaphoric nature of our descriptions of nuclear technologies and by doing so confront and learn from the essential uncertainty they inscribe.

Oppenheimer identified such deference as a matter of style, and the tools for understanding and applying style are the particular expertise of the humanities. If for no other reason than our survival, we must admit the importance of style in "Nuclear Studies"—both in our analyses of the ways nuclear technologies have been made manifest in our society, and in our explorations of the limitations and possibilities of future representations—and we must accept the responsibility of applying ourselves to these studies despite, or rather because of, our uncertainties.

Notes

[1] Cited in Richard Polenberg, *In the Matter of J. Robert Oppenheimer: The Secuirty Clearance Hearing* (Ithaca, New York: Cornell University Press, 2002), 362.

[2] *The Day After Trinity: J. Robert Oppenheimer and the Atomic Bomb*, directed by Jon Else (San Jose, CA: KTEH, 1981), 1:22:10, DVD.

[3] Cited in Mark Wolverton, *A Life in Twilight: The Final Years of J. Robert Oppenheimer* (New YorK, NY: St. Marten's Press, 2008), 267.

[4] I will resist the impulse to refer to scholars in the humanities as "Sith" and instead use the term "humanists," hoping readers will forgive the use of a broad term in this limited sense.

[5] J. Robert Oppenheimer, "The Open Mind," in *The Open Mind* (New York: Simon and Schuster, 1955), 54.

Introduction

The Silence of Fallout

Michael J. Blouin, Morgan Shipley and Jack Taylor

> How does a people react when the entire basis of its existence is fundamentally altered? Most such changes occur gradually; they are more discernible to historians than to the individuals living through them. The nuclear era was different. It burst upon the world with terrifying suddenness. From the earliest moments, the American people recognized that things would never be the same again.
> —Paul Boyer, *By the Bomb's Early Light*[1]

It begins with a moment of silence. The English class, comprised of eighteen and nineteen-year old students, happens to be reading John Hersey's *Hiroshima* (1946) when a 9.0 earthquake strikes the coast of Japan. The Fukushima Daiichi nuclear plant quickly declares a state of emergency. In the United States, broadcasts blend images of utter devastation with references to the atomic bombs and its horrific aftermath. The discourse would all feel a bit anachronistic—if it was not suddenly so fresh. The only activity we can conceive of undertaking with the class is to play a slideshow of images, first from Hiroshima and then from the tsunami-ravaged Japan of 2011. Together we struggle to contemplate, without words, the unsettled clashing of worlds before us: past and present, U.S. and Japan, classroom and "real world." And behind the screen remains the rather formless concept of the nuclear.

How should we discuss nuclear concerns in a humanities classroom? Where is the common ground, the shared site of recognition? After all, most of these students were born after the Cold War's (theoretical) closure. And many of those who were alive during the period seem to have grown complacent. As Aaron Rosenberg observes, "We risk considering ourselves competent with the total nuclear threat simply by virtue of living in a 'post-apocalyptic' time." To approach these vital questions, we might first ask: what happened to Nuclear Criticism, the brief moment when theorists appeared to establish a bond, however unstable, between scholars interested in analyzing cultural phenomena and the nuclear fears of the general populace? Impromptu efforts to explain this juncture for the English class feel unconvincing and consequently fall flat; once more, there is silence. This collection thus aims to re-articulate Nuclear Criticism amidst the continually shifting paradigms of the contemporary landscape.

Nuclear Criticism: A Brief Overview

Nuclear Criticism was a trend in literary theory that emerged most prominently in the 1980s. Many believe it started to coalesce with the 1984 conference at Cornell on the subject, featuring Derrida's seminal discussion of the subject. However, it must be noted that there were a number of important texts that preceded this conference which greatly influenced the trend.[2] This group of theorists had a myriad of goals, though Nuclear Criticism never seemed to agree upon a cogent set of scholarly ambitions. Rather, the trend radiated outward, fading by the mid-1990s, figuratively atomic in its diffusion. Assembled from presentations at the Cornell conference, the introduction to a subsequent collection reads:

> (Nuclear Criticism) arises, on the one hand, out of reading a certain amount of recent criticism and critical theory and feeling that without exception it recounts an allegory of nuclear survival; and, on the other, out of the sense that critical theory ought to be making a more important contribution to the public discussion of nuclear issues... the purpose of uncovering the unknown shapes of our conscious nuclear fears... the use value [...] from predicting the end of things.[3]

The Reagan administration's heightened nuclear rhetoric at this time gave impetus for academics to insert their expertise into discourse on the subject.[4] By way of a reaction, scholars at the conference attempted to outline the primary assertion of Nuclear Criticism as a whole: theorists, already entrenched in nuclear issues on a philosophical level, ought to contribute to discourse on the subject. The approach, on the surface, would be one of *reconciliation*, an undercurrent of trying to bring the humanities, and theory specifically, into the realm of the "real world," of the practical—and urgent—matters plaguing the citizens of a nuclear age. Indeed, as Jeff Smith writes in the opening chapter of the volume, "nostalgia is a means for making sense of experience by putting a satisfying construction on it, plotting it out with narratives and hierarchies and systems of value. If one role of the critic is to undo such constructions, that job is as important now as it was during the Cold War, or ever."

The most widely-cited piece from this conference remains unquestionably Derrida's "No Apocalypse, Not Now (Full Speed Ahead, Seven Missiles, Seven Missives)."[5] The essay extends his deconstructive ethos into the arena of nuclear rhetoric. Therefore, while the goal of the conference was at first to construct a bridge between academia and the nuclear concerns of the populace, Derrida's work pushes back upon Nuclear Criticism with the oppositional assertion that an imagined event only recapitulates the need to break down links between signifiers and transcendent signifieds. He famously writes that the nuclear event is "fabulously textual... a nuclear war has not taken place: one can only talk and write about it... it is a non-event."[6] Derrida argues that any attempt to "ground" the discussion would be folly; words are based in nothing but pure fantasy, caught in Saussure's "endless chain of signifiers." He proceeds to examine how language and the nuclear age are of the same post-structuralist moment: "Deconstruction belongs to the nuclear age... literature has always belonged to the nuclear epoch."[7] Language, he points out, has itself always been in a process of "stockpiling" (more words, more "meaning"). As with the arms race, language is based on speed, the rapid build-up of hasty metaphysical assertions that, deep down, language makes sense. As missiles are stockpiled, so too are the words supporting them, a language designed to re-assure audiences that their existence is "just." Derrida's task, as he lays it out, and the task he calls upon Nuclear Criticism to perform, is to slow down this acceleration, rather than add to it by simply heaping more words with assumed meaning onto the pile: "The critical slowdown may thus be as critical as the critical acceleration."[8] Theorists must not succumb to the need for traditional rhetorical means to match the speed of the political/material movements. Nuclear Criticism, if it is to succeed, must instead continue to breakdown all discourses, regardless of the presumed repercussions in the "real world."[9]

Those responding to Nuclear Criticism often highlight the perceived impracticality of a Derridean model in a world of genuine threats (and potential victims). For some critics, it was the final straw, seen as a moment of absurd critical over-reach by the "cult of high theory."[10] Roger Luckhurst notes: "Even if Nuclear Criticism is not reducible to the universitas, it must still pay attention to the operational effectivity of frames, of framing institutions."[11] K.K. Ruthven's overview entitled *Nuclear Criticism* echoes these concerns: "(Nuclear Criticism) will have to be presented as offering a way out of the impasse of post-structuralism by reintroducing the question of *value*."[12] Meaning must be infinitely illegible and irresistibly legible at the very same moment. This crisis, Ruthven states, stands as the fundamental one facing Nuclear Criticism: "It becomes more important than ever to preserve the nuclear referent and *to resist efforts to textualize it out of existence.*"[13] In an attempt to reconcile meaning with the word, signified with signifier, historians such as Spencer W. Weart likewise re-claim interpretive abilities to scrutinize the nuclear age, a branch of Nuclear Criticism content to return to structuralism for its answers.

J. Fischer Solomon's *Discourse and Reference in the Nuclear Age* (1988), also driven by the exigency of the nuclear age, works to revisit the foundations of literary criticism in just these terms: "The nuclear referent does present a challenge to criticism, an epistemological challenge to think through the consequence of our general textualization of critical knowledge, our unrelenting deconstruction of the referent, of the believe in a physical world whose behavioral properties and dispositions can be objectively calculated and known."[14] Solomon suggests that critics can illuminate "potentialist metaphysics," the probability of a text corresponding to reality. Through meticulous work as a literary critic, they could arrive not at a "hard conclusion" (Solomon admits such reified concepts are philosophically impossible), but at a *likelihood*, a shared understanding of the "structural regularities" of language.[15] After all, he reminds us, is not the purpose of theory to move closer to connection, to a "better understanding"? Solomon contends that we might trust language again, not in a naive or negligent way but in a fashion that starts once more, in earnest, an analysis of words and their relation to our lives. Bradley Fest accordingly acknowledges that we "cannot ignore the imaginative and historical forces produced by the continued dialogue between information and military technologies, between the archive and the Bomb, between the decentralization of the first nuclear age and the networked distribution of the second age in which the nuclear referent has dispersed in a variety of ways."

Written in the immediate aftermath of the Cold War, Peter Schwenger's *Letter Bomb: Nuclear Holocaust and the Exploding Word* (1992) also leaves room for tentative connections to the "real world" by innovative scholars, individuals who did not wish to abandon the advances made by post-structuralism but who longed for constructive approaches to problems that they believed would not be solved by deconstruction alone. These innovative scholars, though largely lost in the wave of "high theory" that persisted through the 1980s, must be re-examined by those who strive to understand the fate of Nuclear Criticism in a post-Cold War world.

Schwenger labors to resurrect a discourse moving multi-directionally, a productive back-and-forth not unlike the politicized notion of deterrence at the heart of nuclear politics: "Launched missiles will not wait for us to finish our sentences, sentencing ourselves to death. Rather the missiles must be returned to their senders even before they are posted, and something in the *timbre* of the sender's mind must make it impossible for them ever to be posted."[16] He, like Solomon and Derrida before him, contends that theorists should re-evaluate the fundamental purpose of language. Nuclear Criticism, according to Schwenger, can assist critics in their efforts to locate a language which is deconstructive but concomitantly posits an ethical position regarding nuclear relationships. Like Schwenger, Paul St. Amour's contribution to this volume seeks an ethical position amidst these missiles/missives, utilizing the theoretical writings of queer theory to interrogate "death drives of church, state, archive, and subject" and, by so doing, re-engage with the nuclear in novel and meaningful ways.

While the title of this collection suggests that on a certain level there has been a tendency to remain "silent" in the post-Cold War world, it is essential to temper this suggestion by recognizing scholars who continue to take up questions posed by Nuclear Criticism. Daniel Cordle stresses the value of this line of inquiry: "The challenge for Nuclear Criticism, whether we conceive of it as a specific theoretical approach, or simply as a critical interest in nuclear issues is to mature beyond its Cold War adolescence and find a way to speak to long-term and more subtle manifestations of nuclear culture." Rey Chow's *The Age of the World Target* (2006), for one, criticizes authors that attempt to make "new meaning" after the dropping of the atomic bomb by simply reiterating deterministic, and oppressive, discursive practices. Examining the emergence of area studies following Hiroshima and Nagasaki, she writes, "Language and literature are rather tools with which to hypostatize the targeted areas... and make them more legible, more accessible, and more available for 'our' use."[17] Chow maintains that language has been used to re-institute boundaries in reaction to the Bomb, making "legible" concepts such as the "essence" or "core" of a particular culture. This claim connects to Gillian Brown in "Nuclear Domesticity: Sequence and Survival," as she points out that the linkage between "peace" and "the feminine" which followed the nuclear event, initially giving the impression of work against armament re-installs linguistic barriers, reiterating "a traditional domestic usage of the feminine, the symbolic value of women to the reproduction of culture."[18] This line of argument, we might add, reverberates with Cordle's emphasis on the "leakage" of nuclear consciousness, its subtle and shifting permutations.

Another work drawn to similar sites of "leakage," Akira Lippitt's *Atomic Light (Shadow Optics)* [2005], builds directly upon Derrida's contributions. By locating three "phenomenologies of the inside" (psychoanalysis, cinema, and X-rays), Lippitt reads the nuclear event as an intersection between post-structuralist rhetoric and the historical development of Cold War culture. The impulse of the twentieth century, according to Lippit, has been one of "hypervisibility," the need to "atomize" any reified images and thus make the legible simultaneously illegible: "Global visibility: a universal archive, in which everything in the world is visible, and everything is visible in the world... reconciling the depth of the body, its volume, with the flatness of the image."[19] His position is not pessimistic, however; he views the notion of the Bomb as a sort of doorway, an ideal moment for shifting the perceived impenetrability of human skin onto a cinematic screen and opening the spectator to the freedoms which accompany figurative atomic dissolution.[20] In chapter five, Joseph Dewey, moving deftly between *The Sopranos* and McCarthy's *The Road*, ultimately concludes that "it is the difficult affirmation of the texts that make up the canon of Nuclear Criticism, a vast body of wisdom literature that reaches beyond any single traumatic historic event. It extends that complicated hope to a humanity that is in perpetual crisis, terrified of the very sense of mystery that alone provides a doomed world its dimension, its nuance, its genuine shock and awe." Conceptualizing annihilation, many of these authors continue to suggest, is never far from conceptualizing emancipation.

It may be from these traversals of disciplinary bounds—from theory to cultural studies to film studies and beyond—that we can recognize a significant future for Nuclear Criticism. John Canaday's *The Nuclear Muse: Literature, Physics, and the First Atomic Bombs* (2000) examines issues emerging from literary theory in conversation with the language of physics. Canaday interweaves the study of literary form with the rhetorical forms that rose up around quantum physics and the development of nuclear weapons. The revelation that scientific perspectives also confront a sense that reality can only be mediated through an "arbitrary, contingent set of partitions" offers an avenue for a regeneration of Nuclear Criticism (Canaday continues his argument in this volume). He states, "Nuclear weapons have been constructed in our society not only as *textual* entities but more specifically as *literary* ones... they do not exist for us except insofar as we are able to imagine in language a set of experiences we have never had."[21] In short, as the story of the Bomb is the story of our contemporary world, it is a narrative worthy of careful study from a wide array of perspectives. For example, Julie Williams stresses "the importance of narrative and how the stories we tell about our nuclear past and possible nuclear futures reveal how we as a society deal with the use of nuclear weapons." The perpetual desire to forge plots from this senselessness (scientific and literary) reaches far beyond the confines of reactionary coping mechanisms; it speaks to something deeper within our shared humanity.

In the end, however, chroniclers of literary criticism are likely to locate Nuclear Criticism as a minor blip on the radar screen. It was, for all intents and purposes, subsumed in the wake of Derrida's essay, presented at the very conference during which the field was supposed to be formalized. Of course, as long as there exists a powerful Bomb, and there is no current reason to believe that this will not always be so, there will be debates concerning how to write about such things—the extinction It brings, the promises people exchange in Its presence. Perhaps older methods of Nuclear Criticism faded away because of the prestige the Derridean method carried with it among those invested in the study of cultural forms. Indeed, perhaps Derrida's assessment in 1984 that the Bomb was already part and parcel of his manner of criticism gave the subject the sepia tones of dated thinking. Yet this collection locates Nuclear Criticism as one possible answer in the quest for a theory more grounded in "real world" concerns, having produced some of the most innovative, inspired readings of the post-war period. By compiling these essays, we affirm that both established and emergent scholars continue to attend to these issues, even if the issues manifest themselves in entirely unexpected places. Though engulfed by deconstructive "explosions," these innovative approaches lead us back to the rubble of language with eternal hope to construct something more beautiful.

Geopolitics, the "Real World," and Nuclear Criticism

In the immediate years following the Cornell conference, Nuclear Criticism, as a nascent field of inquiry, ebbed and waned and finally lost its critical footing. Such a narrative of decline speaks to the difficulties of locating a single method adopted by Nuclear Critics and, more distinctly, to the ubiquity of the nuclear. While it would be short-sighted to suggest that all critical studies of the nuclear simply faded with the declared end of the Cold War, we cannot understate the innocuous position that the nuclear now appears to hold as a consistently diffused, and diffusing, referent. Moreover, while the nuclear never found an institutionalized home in the academy, against Derrida's prediction, it leaks out and contaminates discourses of all kinds. This problematic emerges cogently within everyday life, where the nuclear continually arises as a fantastic, but fully present, threat capable of negating humanity. Yet familiar paradigms continue to be applied to around the topic, refusing to acknowledge the slow drift, to borrow from Jonathan Schell's recent work, "toward what some have termed 'nuclear anarchy'," the growing reality that the Cold War did not end the nuclear, but rather made it an operational norm.[22]

By making the nuclear—in all its guises, from weaponry to energy— an accepted practice (at least, for the nations deemed "responsible enough"), proliferation dictates a geopolitical landscape trapped between the potentials of nuclear energy and the apocalyptic tropes that continue to constitute it in the public imagination. For Schell, this problematic became astute during the post-9/11 era, an era shaped in the United States by the Bush Doctrine and, more specifically, its policy of renewing the nuclear as a common referent and a method of deterrence. Indeed, in the years following 9/11, the nuclear persistently emerges, seeping through the cracks of discourse, illustrating the many ways in which the "real world" continues to exist amongst, and in many ways is defined by, silent specters of the nuclear.

On October 7, 2002, in his speech outlining the Iraqi Threat, President George W. Bush signified and energized the threat of the nuclear by honing in on a common trope: "Facing clear evidence of peril, we cannot wait for the final proof—the smoking gun—that could come in the form of a mushroom cloud." And: "Understanding the threats of our time, knowing the designs and deceptions of the Iraqi regime, we have every reason to assume the worst, and we have an urgent duty to prevent the worst from occurring."[23] As Bush's language suggests, apocalyptic concerns, specifically concerns referencing the nuclear, offered a rhetorical device for garnering mass support through mass hysteria. In connecting the nuclear to the "terrorist," the Bush administration directly challenged the perceived vulnerability of key American values in light of expanding threats decentralized from the nation-state. Subsequently, we witness how the nuclear endures, wielding the bulk of its power as a mere threat, as mere potentiality, anchoring a geopolitical landscape in which the nuclear has effectively taken up residence everywhere, at all times.

Less than fifteen years after the fall of the Soviet Union, the Bush Doctrine, as Schell outlines in *The Seventh Decade* (2007), reinvigorated this potentiality, while simultaneously demonstrating its proliferation as a means to manifest and mobilize outright imperial tactics. In key moments of self-contradiction, the nuclear becomes the source of domestic peril and, simultaneously, the solution, the great equalizer. Little, subsequently, has changed since the Cold War; and without critical voices to interrogate the application of weapons of mass destruction, we might soon exist in a global world defined by a "unique riddle of the vacillating, intermittent, and currently stalled human encounter, now more than sixty years old, with what is still the only technology that can put an end to all human beings."[24] According to popular thought, the nuclear, rather than diffusing, evolved into a rogue tool. On the one hand, it justifies and legitimates the United States as an imperial mediator, dictating who can (and cannot) implement life-destroying, and life-maintaining, technology. On the other hand, as opposed to two nations—the United States and Soviet Union—having their hands hovering over the proverbial red detonator, the threat of the nuclear, as Bush's speech indicates, becomes detached from particular nation-states that, at least discursively, predicate acceptable levels of force through "just" methods of war. The nuclear concurrently supplied Bush's tool for battling terrorism and an almost universalized technology capable of producing civilization-ending weaponry. As with the Cold War, the nuclear danger became, according to Schell, "an axle around which the wheel of geopolitical events is turning."[25] Given recent global events, there is no clear end in sight.

In truth, over the past two years, the nuclear has re-emerged as a defining trope in political rhetoric and material application. From Iran to North Korea, the U.S. positions itself politically in accordance to the perceived nuclear capabilities of these "rogue" states. Rhetorically, rather than through concrete statements or plans of use, the nuclear functions as a discursive device, utilized for political leverage to invoke terror and, more insidiously, justify pre-emptive international sanctions or direct unilateral confrontation. As recent as May 8, 2012, Vice President Joe Biden vehemently challenged the Iranian government, warning that a timeframe for peacefully resolving the nuclear standoff was closing "in the near term."[26] Almost directly echoing the Bush Doctrine, Biden speaks to the heightened problematic of controlling the nuclear *through the threat of the nuclear*. Such a situation, however, is not solely unique to the United States. Indeed, as Pakistan and North Korea have illustrated, the nuclear persists as an operational ethos for maintaining sovereign power. The Committee for the Peaceful Reunification of Korea directly challenged a recent U.S. declaration for Pyongyang to relinquish its nuclear development programs as a "grave provocation."[27] Yet, as with Biden's comments, does this "grave provocation" simply function as rhetorical fallout, revealing the growing divide between discussions of nuclear capacity—including the capacity of safer and more efficient energy—and the application of nuclear weaponry?

Indeed, as Nuclear Criticism repeatedly locates, the very problem of the nuclear often revolves around this dialectic, exposing holes within the nuclear yet remaining exposed by its seemingly endless dispersal (the true meaning, for Derrida, of an *epoché*). And, as the United States recently demonstrated, such a dialectic allows the nuclear to operate as an apocalyptic image, signifying the potential destruction of life, while promoting a humanist "ethics" maintained, ultimately, by nuclear proliferation. This can been seen in the recent dismantling of the B-53 nuclear bomb, the largest and most powerful nuclear weapon to date, 750 times as powerful (9,000 kilotons) as the bomb dropped on the city of Hiroshima in 1945 (12 kilotons). But this dismantling, like the nuclear trope in general, is a mere mirage, veiling the continued proliferation that exists simultaneously as a threat, a method of deterrence, and a unifier of people and places. Indeed, while one B-53 bomb is now decommissioned, it is estimated that the U.S. still has 1,800 strategic warheads capable of being launched by land or sea in twelve minutes—and another 2,500 in reserve.

This problem—between the image of the nuclear versus the reality of its application—returns us to the role of Nuclear Criticism. In truth, with the cancerous proliferation of nuclear capacities, exacerbated by political rhetoric, Nuclear Criticism as a tool of the humanities can and should target the "real world" as its site of interrogation. In chapter ten, Patrick Sharp explores *The Hunger Games* to stress that "in the end, Suzanne Collins provides the hopeful thought that we might be able to evolve quickly enough to avoid extinction via nuclear and ecological apocalypse. Her work shows that nuclear weapons have not been forgotten… By foregrounding the toxicity of nuclear weapons, *The Hunger Games* trilogy continues the tradition of […] contextualizing the dangers that nuclear technologies pose to all life." Sharp forces the reader to consider the price of ignoring nuclear narratives. Thus, in the same moment in which the United States deconstructs the largest nuclear weapon, it is essential to ask: where will we continue to find critics deconstructing the various meaning(s) of this act, specifically given the remaining arsenal and the seemingly daily reminders of nuclear scare tactics as nations fight to control the spread of nuclear capacity? Moreover, beyond the sublime spectacle of the Bomb, the Fukushima Daiichi nuclear plant failure unveils a heightened level of anxiety that re-energizes post-Chernobyl phantasms. The anxiety over fallout, the ways in which the nuclear remains viral, exacerbates the temporal positioning of the nuclear, re-activating the fears that defined the original nuclear epoch. Echoed in other disasters—from fears of nuclear failure in the United States following an east-coast earthquake to the apocalyptic language that accompanied the BP Oil Spill and other similar human-made disasters—the nuclear continues to function as a signifier of political, environmental, and personal apocalypse.

To be sure, the issues surrounding the nuclear have not been solved and decoding its continued diffusion requires more than a policy of deterrence and more than scientific discourses that anticipate a definitive, empirical answer. As Derrida originally argued, we cannot approach the nuclear through antiquated metaphysical assertions; however, as critics have been apt to point out since Derrida's presentation at Cornell, neither can we ignore the concreteness of nuclear weapons, specifically within a contemporary moment removed from critical interrogation or centralized oversight. This problematic complicates any clear analysis, but it also speaks to the limits—and potentiality—of theorists capable of assuming the mantle of both positions simultaneously, to maintain a type of cognitive dissonance that forces the reader to think about discursive work *and* the material application of nuclear technology. Nuclear Criticism provides guidance by constantly renewing the conversation, allowing the nuclear to exist neither as a naturalized political tool nor solely as an object for quiet contemplation.

Framing *The Silence of Fallout*

This volume explores various contemporary manifestations of nuclear anxiety and advocacy as well as the periodic gaps where critical engagement seems to grow inaudible. Mark Pedretti expresses the treacherous conditions for this re-engagement, noting that "if we are to align a synchronic periodizing concept (style) with a diachronic one (historical event), as the Nuclear Critics have done, then we are left with little alternative but to force impossible stylistic constraints onto texts by vesture of their historical location." With these constraints ever-present, one must pause to ask what the role of the humanities is during times of heightened nuclear anxiety as opposed to times of decreased concern. How has this changed over the last fifty years? How do we remember nuclear catastrophe and forecast its potential devastation as actual witnesses, and citizens formed in the Cold War era, fall victim to the passage of time? What must be preserved—and what must be forgotten? As anxieties often manifest in different channels of representation, what is the relationship between the "nuclear threat" and the "terrorist threat" post 9-11? These are several of the pressing questions investigated within this collection.

Our framework places scholars who were active in earlier expressions of Nuclear Criticism in conversation with emergent scholars who are striving to negotiate the field moving forward. The collection therefore, as a whole, synthesizes around dialogic moments of agreement and departure. As political paradigms shift, and awareness of nuclear issues concomitantly manifests in alternative forms, the collection establishes groundwork for the next generation of individuals that will struggle to come to terms with the innumerable legacies of the nuclear.

Few works have taken a step back to survey the role of scholars or the state of Nuclear Criticism as a loosely-integrated enterprise (notable exceptions are K.K. Ruthven's *Nuclear Criticism* and the short-lived journal, published from 1988 to 1995, entitled *Nuclear Texts and Contexts*). Even fewer works offer insight into the years following the heated rhetoric of the Cold War. This collection asks where Nuclear Criticism stands today. As there is no sign of a world freed from this crisis, where will outlets remain for critical thinking on the subject? William Knoblauch asserts, in his contribution to this volume, "Nuclear fear, a pervasive reality of the Cold War, now seems to be a relic of the past; unfortunately, the nuclear threat remains very real." The volume examines recent discourse in order to contemplate the future of Nuclear Criticism, a conversation we believe to be long overdue.

This dialogue must first address how the notion of the nuclear has changed over the past two decades and how recent cultural works re-orient our relationship to these issues (from *The Road* to the evolution of video games). In analyzing these transformations, contributors to this collection repeatedly locate disjunctive moments in which eschatological concerns disguise radical re-assessments of the present. Darwinian paradigms of "progress" are deconstructed in order to forge productive avenues of inquiry: the death drive of the State and its relationship to queer theories of temporality, the emergence of "secondary circuits" in literature that foster a "nuclear uncanny," the role of the internet as a ubiquitous archive formed in part by the logic of impending nuclear disaster. Each of these original readings shares the ambition to shape from the remnants of Nuclear Criticism not a story of chaos or dread, but a tale from which—as Jessica Hurley eloquently writes—we may still learn "how to stop waiting." This volume therefore articulates an ethical agenda for the next wave of Nuclear Critics. It does so by shifting away from constructions of an apocalyptic future, as well as a nostalgic past, and instead examines how the nuclear—both literally and figuratively—exists within us, self-enclosed in the post-Cold War here and now. For better or worse, it is a part of us.

Our title, *The Silence of Fallout*, refers to a lack of conversation in our English classroom on that fateful day, as we stared collectively alongside Hersey at images of the disasters, but lacked sufficient words to express our rage, our guilt, our empathy. It simultaneously cites a shrinking number of academic voices engaged in discussion regarding the proliferation of nuclear arsenals. Perhaps we have been listening in the wrong places, succumbing to the banalities of government discourse and the predictability of Hollywood formulas. Nuclear issues will likely never disappear, be they material or speculative in nature, and as long as they persist, we must continue to converse. A sense of hope remains in the pages to come; it remains in efforts to slow the conversation down, to delay the last word on the matter. Our alternative—deafening silence—is simply too terrible to dwell upon. And if we ought to speak with one another, why not today?

Notes

[1] Paul Boyer, *By the Bomb's Early Light: American Thought and Culture at the Dawn of the Atomic Age* (Chapel Hill, NC: The University of North Carolina Press, 1994), 3-4.
[2] See the works of Karl Jaspers, *The Atom Bomb and the Future of Man*, trans. E.B. Ashton (Chicago: University of Chicago Press, 1963), Helen Caldicott, *Nuclear Power is Not the Answer* (New York: The New Press, 2007), and Norman Cousins, *Modern Man is Obsolete* (New York: Viking Press, 1945), for a sampling of the early cultural commentators on the nuclear question.
[3] Richard Klein, "Introduction," *diacritics* 14, no. 2 (1984): 2.
[4] For more on the link between the "official" rhetoric of the American government during the nuclear age and the problem of referentiality, see Stephen Hilgartner, Richard C. Bell, and Rory O'Connor, *Nukespeak: The Selling of Nuclear Technology in America* (New York: Penguin, 1983). In this work, the authors examine how government officials attempted to stop the de-stabilization of language that came with the Bomb by insisting upon over-use of codes. This overuse was designed to restore a sense of balance between the Word (the code) and power over the "meaning" (the "atomic secret") by government narrators. This performance often resulted in little more than linguistic games (a critique that could be leveled at much of Nuclear Criticism as well): "The popular image of an atomic secret as some diagram or formula on a single sheet of paper has little to do with the real secrets that remain in the world of nuclear weapons design... it is not the lack of secret information that stops nations from building hydrogen bombs" (67).
[5] Lest they be forgotten, many of the other contributions at the Cornell conference actually attempt to restore linguistic or metaphysical "meanings." As just one example, Mary Ann Caws offers an alternative which channels a mystical return to

the "rhythms" of language, an appreciation of the communal experiences of discourse. She states, "No holding together impulse is laughable, not even our desperate attempts at conversing and at singing as we can." In Mary Ann Caws, "Singing in Another Key: Surrealism Through a Feminist Eye," *diacritics* 14, no. 2 (1984): 70.

[6] Jacques Derrida, "No Apocalypse, Not Now: Full Speed Ahead, Seven Missiles, Seven Missives," *diacritics* 14, no. 2 (1984): 23.

[7] Ibid., 27.

[8] Ibid., 21.

[9] Jean Baudrillard is equally ludic with the imagined nuclear event as a critical trope. He posits the figurative explosion as the beginning (and end) of language's undoing: "Every political, historical and cultural fact possesses a kinetic energy which wrenches it from its own space and propels it into a hyperspace where, since it will never return, it loses all meaning... the particle accelerator which has smashed the referential... every fact becomes atomic, nuclear, and pursues its trajectory into the void." In Jean Baudrillard, *The Illusion of an End*, trans. Chris Turner (Stanford: Stanford University Press, 1992), 2.

[10] See Christopher Norris, *Uncritical Theory: Postmodernism, Intellectuals and the Gulf War* (Amherst: University of Massachusetts Press, 1992).

[11] Robert Luckhurst, "Nuclear Criticism: Anachronism and Anachorism," *diacritics* 23 (1993): 92.

[12] K.K. Ruthven, *Nuclear Criticism* (Melbourne: Melbourne University Press, 1993), 27, emphasis ours.

[13] Ibid., 74, emphasis ours.

[14] J. Fisher Solomon, *Discourse and Reference in the Nuclear Age* (Norman: University of Oklahoma Press, 1988), 31.

[15] Ibid., 81.

[16] Peter Schwenger, *Letter Bomb: Nuclear Holocaust and the Exploding Word* (Baltimore: The Johns Hopkins University Press, 1992), 20.

[17] Rey Chow, *The Age of the World Target: Self-Referentiality in War, Theory, and Comparative Work* (Durham: Duke University Press, 2006), 15.

[18] Gillian Brown, "Nuclear Domesticity: Sequence and Survival," In *Arms and the Woman: War, Gender, and Literary Representation*, ed. Helen M. Cooper (Chapel Hill: University of North Carolina Press, 1989), 287.

[19] Akira Lippitt, *Atomic Light (Shadow Optics)* (Minneapolis: University of Minnesota, 2005), 47.

[20] Maurice Blanchot shares this utopian vision of the nuclear event in *The Writing of the Disaster*, trans. Ann Smock (Lincoln: University of Nebraska Press, 1980). For Blanchot, language, in the face of the bombing, reveals a tremendous gift: release from mastery, from the teleological underpinnings of History.

[21] John Canaday, *The Nuclear Muse: Literature, Physics, and the First Atomic Bomb* (Madison: University of Wisconsin Press, 2000), 222.

[22] Jonathan Schell, *The Seventh Decade: The New Shape of Nuclear Danger* (New York: Metropolitan Books, 2007), 14. See also Jonathan Schell, *The Fate of the Earth* (New York: Avon Books, 1982).

[23] President George W. Bush, "Address Regarding Iraq" (speech, Cincinnati, OH, October 7, 2002), accessed Jun 2, 2012, http://www.johnstonarchives.net/terrorism/bushiraq.html.
[24] Schell, *The Seventh Decade*, 11.
[25] Ibid., 6.
[26] Olivier Knox, "Biden: Time running out for diplomatic resolution of Iran nuclear standoff," *ABC News*, 8 May 2012, accessed June 2, 2012, http://abcnews.go.com/Politics/OTUS/biden-time-running-diplomatic-resolution-iran-nuclear-standoff/story?id=16305379#.UMfFlXc8CSo.
[27] Associated Press, "North Korea threatens to bolster its nuclear arsenal amid worries about another atomic test," *The Washington Post*, 10 May 2012, accessed May 12, 2012, http://www.washingtonpost.com/world/asia_pacific/north-korea-threatens-to-bolster-its-nuclear-arsenal-amid-worries-about-another-atomic-test/2012/05/10/gIQALS86EU_story.html.

CHAPTER ONE

"WHAT WORKS":
INSTRUMENTALISM, IDEOLOGY,
AND NOSTALGIA
IN POST-COLD WAR CULTURE

JEFF SMITH

War is something that people make, and therefore is embedded in culture. What first got me interested in Nuclear Criticism, back before we knew it as that, was a sense that this fact had been overlooked. The early 1980s were grim days of re-escalating Cold War, and there was no shortage of worried efforts to explain the nuclear threat and its implications. Most of these, though, struck me as heading in one of two wrong directions. The first was a version of the widespread, seemingly common-sense idea that the "Nuclear Age" was a sudden new development, a rupture in history that had fundamentally remade the world since 1945. Competing great powers will always arm themselves as lethally as they can, so once the secrets of the atom were known, that necessarily meant building lots of nuclear bombs. The nuclear threat, then, had no particular cause, other than the autonomous advance of science. But it had huge consequences, changing not just war and world politics but, perhaps, the human condition itself. In one ambitious and widely read exposition of this view, Jonathan Schell argued in his 1982 *The Fate of the Earth* that human life depends for its meaning on having a future, and therefore the chance of sudden nuclear destruction had cast a pall over every human enterprise.[1]

Mirroring this view was a different, almost opposite theory in which the psychological disorder came first. The nuclear threat arose from original sin, or a secular counterpart—aggressiveness, territoriality, a need for enemies, a fetish for things that go "boom"—that reaches back to time immemorial. The human condition had not recently changed; in fact, that was the problem. Nuclear weapons are madness, but war itself is madness, and humans have always been mad. They suffer from an ingrained paranoia, a "Neanderthal mentality" that once gave us stone clubs and now gives us ballistic missiles. In this view, the nuclear threat did have a cause. But a cure? There is no therapy for human nature, no solution to original sin on this plane of existence.

Both these twin fatalisms seemed at odds with some obvious facts. Most people were going about their daily routines more or less as they always had, with no sense that life had recently lost meaning. (Plus, the idea that these are the Last Days was at least as old as Christianity.) And while human nature is unquestionably flawed and fallen, it is also suffused with hopes and ideals. Most people mean well and want the best, and any explanation of the nuclear crisis that neglected this was missing at least half the story. So, other than "better dead than Red," what way of thinking had led large numbers of people to regard so terribly dangerous a situation as acceptable, maybe even as a positive good? How and when, specifically, did that way of thinking arise, and how might it change?

It was characteristic of much antinuclear writing that it would lapse at this point into the "We" fallacy, as I called it in my 1989 book *Unthinking the Unthinkable.* Speaking on behalf of a "we" or "us" is a standard writer's device, but here it became a fallacy by smudging over the question of agency. Readers were being told that "we" are high-tech Neanderthals, that "we" have "all" collectively willed the nuclear threat into existence. It shows how unyielding the arms race had come to seem that even writers who obviously didn't imagine themselves in that "we," and who were sometimes willing to name particular culprits—Ronald Reagan, Edward Teller, the military-industrial complex—would still conclude by instructing "us" to overcome "our" neurotic need for war and violence. I was not the first critic to point out that if everyone is at fault for something, then as a practical matter no one is.[2] Demands that "we must change" are exhortations, not solutions, and are likely to reach only those who are already convinced.

But this wasn't merely a case of preaching to the choir. The "We" fallacy had the effect of depoliticizing the nuclear problem, of obscuring the ways in which it represented a contest of forces and clash of agendas. It is true and important, as we will see in a moment, that a national community is a "we" for some purposes. But "we must change" confuses "we" the writer and sympathetic readers with "we" the people at large, or worse, with "we" the nation as actually led by people who have no intention of changing. It's a polite way of not saying "*they* must change," even when that's what the writer actually means. How, after all, are we on the antinuclear side going to get them to change? Especially if they're Neanderthals? What power do we really have? This is the conundrum that follows from placing nuclear weapons beyond the pale, from ascribing them not to particular political developments but to the human predicament itself. Solving a problem politically means getting specific, locating it in the stream of history, where there is no one predicament but a series of them. Some of these are long-lived, but none is permanent. The goal of criticism should be to separate out the elements, to identify the alignments of ideas and opposed forces, that make each problem the particular problem that it is.

To do that, I believed, we (critics) needed the humanities—the disciplines devoted to careful study of cultures, their histories, their texts and imaginative products. Like any human creation, the nuclear threat had cultural origins, which is to say it neither appeared suddenly in 1945 nor dates back to the dawn of time. Some arrangements *seem* timeless because cultures are slow to change; the relevant time scale is not years or decades, but generations and centuries. If anything, the speed of change starting in 1989 took many antinuclear writers, including me, by surprise. We seem in many ways to be living in a different world now, one that no longer puts human existence on a hair trigger. But our post-Cold War experience, so far, still extends only to years and decades, a scale on which cultures are more continuous than not. The nuclear threat itself has dimmed but not disappeared, and we've had no end of other worries and alarms, especially since September 11th, 2001. In what follows, I would like to review some of the lessons I believe I learned from Nuclear Criticism—from close cultural, textual and humanistic analysis of the nuclear threat—that might be helpful in understanding our present preoccupations, especially terror, torture, proliferation, nation-building and "failed states."

Instrumentalism

Of course, there are actual "we's" that operate politically in the world. A very important one is the nation-state—an agglomeration of people (or peoples) who understand themselves to be joined together for purposes of acting through the agency of a single government. American writers who speak of "us" as doing this or that are typically referring to citizens of the United States, whose relatively wealthy and stable government does give its people considerable power to act jointly. Again, it's important not to reduce large, complex societies to a simple "us" versus "them," the indispensable rhetoric of those who would encourage and justify war. But it's nonetheless true that much of what has made the world of the last few centuries function as it does is the arrangement of peoples into (usually) coherent nation-states. The "We" fallacy is least fallacious when it expresses this, when it is used to acknowledge our own or others' historical agency as exercised through "imagined communities," in the title phrase of Benedict Anderson's influential study of nation-forming.[3]

As Anderson and others have detailed, the making of such communities was a cultural change that can be specifically located in history. It happened at different times in different places (and is still happening, or failing to, in some regions), but in Western Europe, nation-states were largely products of the period now known as "early modern," the sixteenth and seventeenth centuries. Shakespeare, a highly perceptive observer who lived in one of those developing states, saw the change taking place and artfully captured it in his dramas. I found these, therefore, very helpful in understanding the nuclear crisis. Shakespeare and his contemporaries were worried about many of the same things that worry us today, including the new powers represented in new ways of making war. *Henry V,* Shakespeare's docudrama (as we might now call it) of the English conquest of northern France two hundred years earlier, is full of moral doubts and demurrers—remarkably so for a play celebrating the great victories of a national hero. On the eve of battle, King Henry, in disguise, finds himself debating his own skeptical soldiers over questions of war guilt. "I am afeard there are few die well that die in a battle," says one, "for how can they charitably dispose of any thing, when blood is their argument?" Henry has various answers, but the most important is the war's socially improving capacity to dissolve old divisions of region and class. Henry's army is captained by English, Scottish, Welsh and Irish officers, historic adversaries who learn to cooperate for the cause. And in one of Shakespeare's most famous speeches, Henry inspires his troops with a new vision of social rank: Those who fight alongside him, however humble by birth, will rise beyond their supposed social betters and join a new "we" in which they become equals of the king himself: "We few, we happy few, we band of brothers."[4]

The lesson here is that modern responses to the technologies of war cannot be separated from modern views of war itself and of the states that fight them. These are themselves technologies of a kind: War is an instrument for creating states, and modern states are, or at least claim to be, instruments for advancing the interests of common people. That is why they command loyalty. And the changing means of warfare reflect this. At the famous battle of Agincourt, French knights on horseback fall before Henry's smaller army of yeoman archers—sturdy, plain English commoners with longbows who grind the flashy French aristocrats literally into the mud. Those who make war have always wanted to win, of course, but Henry is unusually focused on effectiveness to the exclusion of other concerns. His willingness to downplay traditional social rank and Christian charity turns out to be a key strategic advantage, allowing him to make even the patchwork, not-quite-United Kingdom into a winning cause—the kind not just worth dying but, importantly, worth *killing* for, in an era when killing was still thought to pose a risk to one's immortal soul.

Gradually, as they consolidated their new power, emerging nation-states rejected those old religious constraints too. This made them moral laws unto themselves—"sovereign" entities of a sort arguably not seen in the West since the Roman Empire.[5] In some cases, the state itself became a new religion of a kind. Modern nations founded on explicitly revolutionary ideas, including the United States and the Soviet Union, would prove especially effective at presenting themselves as great causes, history's agents for advancing "the People" (or "the Workers"). And in America, prosperity added to the state's prestige, while also enabling massive further improvements in destructive technologies. These could now be favorably imagined as instruments of power placed in the hands of (an instrument of) the people. You don't need a theory of mass neurosis to see why the people, for the most part, approved.

The problems with all this, though, are equally obvious and quick to appear. The defeated French at Agincourt complain to Henry that they will need time

> To look our dead, and then to bury them;
> To sort our nobles from our common men.
> For many of our princes—woe the while!—
> Lie drown'd and soak'd in mercenary blood;
> So do our vulgar drench their peasant limbs
> In blood of princes…[6]

The horror isn't just that they lost, or that they're dead, but that they're all mixed together—a kind of awful parody of what it means to be one nation. And this was a problem that the new technologies would only worsen. The arrow storms of Henry's longbowmen were still pre-industrial weapons of comparatively low power. They merely killed people and horses. Far more troubling, both strategically and morally, were weapons that could kill whole cities. Granted, this had always been one possibility in war; Carthage was leveled in 146 BCE, and Troy a thousand years before that. The first "weapon of mass destruction" was fire. But by the late Middle Ages, fire had been harnessed in new and considerably more frightening ways. The image we have today of battlefields as hellscapes of smoke and thunderous explosions dates from the widespread use of cannons, the superweapons of Shakespeare's time.

To medieval observers, early cannons were the handiwork of Satan. Some were even molded to look like devils or dragons.[7] They were forged in fire and heat, and the complex techniques of engineering, chemistry and ballistics they demanded were dark arts, the secret disciplines of specialized guilds that had learned, it seemed, how to summon the forces of hell. Here was real alchemy, an early instance of the demonic knowledge that would recur in the hands of "mad scientists" in so many B-movies of the atomic age. Artillery, moreover, could destroy from afar, joining many other weapons—from ancient poisons to medieval bows to early rockets, mines, submarines, and today's lasers and drone aircraft—that at various times have been condemned as the disreputable methods of cowards and sneaks. They made killing too easy, violating the old ideal of skilled and virtuous combat at sword's length with an equally matched opponent. War should be an "open and manly" contest of chivalry and personal heroics, including "a fair and honourable approach to the enemy."[8] Cannons just blasted away, killing indiscriminately. And in flattening buildings, city walls, castles and princely estates, they also flattened social distinctions. It was not enough that power had shifted from lords and barons to massed ranks of anonymous archers; now even those gave way to a grimy class of low-born gunners who did not really fight at all, but spent most of their time in workshops pounding out machines which, as they became bigger and more powerful, only big, centralized, bureaucratized armies and monarchies could afford.

Already by the late Middle Ages, kings could in effect threaten to "go nuclear" against cities and their civilian populations. Machiavelli, whose cynical view of statecraft was already notorious enough to rate mentions in Shakespeare, had even advised them to do so: "There is no secure mode to possess [cities] other than to ruin them. And whoever becomes patron of a city accustomed to living free and does not destroy it, should expect to be destroyed by it."[9] Shakespeare's Henry is not quite so ruthless. But he threatens to be: "If I begin the battery once again," he warns while aiming his guns at one French town, "I will not leave the half-achieved Harfleur till in her ashes she lie buried." A city that has done nothing specifically wrong is "guilty in defence" if it happens to be in the wrong country, and Henry—the literary model of a *good* king!—declares this sufficient cause for its total destruction. He even makes the modern-sounding suggestion that the "waste and desolation" will be automatic, not so much his or anyone's conscious aim but just the logic of "impious war" itself.[10]

But, just as in the atomic age, two could play the same game. Cities did indeed fight back; the French could build guns too, and in *Henry VI Part 1*, the play that continues the story of *Henry V*, a single shell from one of these kills two English nobleman-generals, both combat heroes, one the former military instructor to Henry V himself. To add to the insult, the gun isn't even fired by a master gunner; it's a mere boy, the French gunner's son and apprentice, who happens to be tending it when he sees his chance for a lucky shot.[11] Like nuclear secrets, the secrets of any terror weapon proliferate. And once war comes to be defined by terror, peace on a victor's terms becomes very hard to maintain.

It is no surprise, then, that *Henry VI*, like the events it chronicles, anticipates many of our more recent anxieties. The parallels are striking: Having listened to counselors who want war for other reasons, and having attacked a country that was not really the enemy, a ruler wins a series of quick victories, declares "mission accomplished" and sets up an occupation. Only then begins the real struggle, a grinding mess that lingers on to baffle that ruler's successors. The occupation is expensive, under-supplied and vulnerable "for want of men and money." Its local allies are unreliable and quick to switch sides. The occupying force is too small, making the troops easy targets for (what we now call) "asymmetrical" attacks—like that shot from a hidden gun, or soldiers sneaking past roadblocks by disguising themselves as harmless peasants.

The insurgency, moreover, gathers force from religious extremism. The antihero and most interesting character in *Henry VI, Part 1* is Joan la Pucelle, the teenage shepherd's daughter who comes from nowhere to take command of the French forces. Joan of Arc, as we know her, is a very unlikely general—wrong age, wrong class, even the wrong gender. But she is certain that God has spoken to her. And she is effective, which, in modern war, is what matters most. What makes her so are qualities that she shares with Henry V: impatience with rank and conventional decorum (she mocks the English for their "tedious" and "silly-stately style"), and a genius for mobilizing the same nationalist urges that had previously served the enemy. Joan grasps that the French are a nation in the making too. Her visions are the beginning of their own imagined community, as her tactics simply carry further the occupiers' innovating spirit, their earlier willingness to embrace the new and unchivalrous as long as it seemed to work. Joan cannot conjure actual devils, but she does conjure her countrymen's nascent national loyalties, and this makes her words "like roaring cannon-shot." In that aptly fearsome image, Shakespeare captures the fusion of the nation's technical instruments with the nation itself.[12]

Needless to say, the English in *Henry VI* find this appalling. On their way to defeat and an ignominious, "effeminate" peace, they complain at length about the "hellish mischief" of "that damned sorceress," and about the French willingness "to join with witches and the help of hell!" instead of politely meeting "in the field" to "take up arms like gentlemen." But they find themselves squeezed from both sides, under pressure both from the continuing power of the old religious values they had underplayed, and from the new, nationalist values and energies that their own occupation has helped ignite. A new nation is emerging under their feet, and the great cause now is to see them ejected from it. The forces that made their conquests possible are thus thrown into reverse, and the blowback—as we now call it—aggravates their "civil dissension," eventually bringing down their own government in civil war.[13] America's misadventures abroad have not brought it down (yet), although the Soviet regime's arguably did. But even in America, the "culture wars" inflamed in part by defeat in Vietnam continue.[14] And in this "cold civil war," as it has been called, we find Americans continuing to struggle, like their opponents if not more so, with the moral contradictions at the heart of the modern state.

Ideology

To emerge as the leading political organization of today's world, the nation-state had to assert itself against—and, where possible, absorb—already existing communities, groups defined by class, religion, region, ethnicity and dialect. And to do that, it had to persuade people to put patriotism, at crucial moments, above the traditional identities and loyalties that had bound them into those other groups. Some states have gone even further and experimented with rooting out the old ways by force. Robespierre's France, Hitler's Germany, Mao's China, and Pol Pot's Kampuchea actively disrupted, even outlawed, people's other affections as a matter of policy. The ambition was to rebuild history from scratch, painting over unapproved thoughts and feelings with the ideology of the state. But except in dystopian fictions, like George Orwell's *Nineteen Eighty-Four* (1949), efforts like these have ultimately failed.[15] Most real states still offer many sub-identities, many other possible objects of ultimate loyalty, some of which come attached to value systems that can trump the state's claim to sovereign authority—moral standards against which the state might still be judged and found wanting. States are, furthermore, bound by what Shakespeare called the "law of nature and of nations."[16] They exist in the real world, where certain facts are true and others aren't, and they all share the planet with other states, a situation that becomes highly unstable if one of them declares itself sovereign judge of its own actions without at least provisionally granting the right of the others to do the same.

All this makes for a complicated picture when applied to the United States. If there is an instinct truly ingrained in people since time immemorial, it is the sense that destroying things and killing or hurting people are wrong. Yet these are easily foreseeable consequences of war. War is therefore presumptively wrong. Since ancient times, this has led to efforts to bring it within rules—first, codes of honor and chivalry, then religious doctrines of "just war," then the rules of modern diplomacy and Enlightenment ideals of human rights. The combined results are enforced, or at least acknowledged, today in military codes of conduct, the Hague and Geneva Conventions, the UN treaty, and other agreements outlawing war crimes and banning certain classes of weapons. In many cases, America took the lead in formulating these rules in their modern forms and preaching them like gospel to the world's other nations.

But the U.S. is also a modern state, meaning that it participates in the instrumentalist bias in favor of what works—the new, modern, pragmatic approach that nation-builders like Shakespeare's Henry pioneered, and against which traditional notions of honor, sin, and justice seem fussy and old-fashioned. Indeed, America more than most nations prides itself on its practicality and technical know-how, which are widely seen as key reasons for its success. It is thought to be more than mere happenstance that the U.S. led the world in harnessing the atom and figuring out how to build nuclear bombs. It has led the world in lots of things. And America's practical successes are, paradoxically, all mixed up with its high ideals, its insistence since its founding on its special, "exceptional" role as a bearer of history's values and a beacon of liberty to the rest of the world. It was America's freedom, its essential goodness, its willingness to throw everything into the fight against evil and tyranny, and even its faithfulness to God—so the thinking goes—that made it both the wealthiest nation in the world and the most powerful.

But if America is exceptional, it is exceptionally disturbing to discover that there is a serious problem. In an image that American writers have turned to for generations, the machine has noisily intruded on the peace of the garden.[17] What counts as "the machine" has changed—at one time, it was that newfangled technology called the railroad—just as what counts as a terror weapon has changed many times since the bygone days of castles and cannons. But the terror is real, and the U.S. played a key role in creating it. In some perverse, unforeseen way, by the late twentieth century American know-how had brought the whole world to the brink of destruction. And not only that: it had left America itself undefended against nuclear missiles, a high technology of America's own design. For a sunny optimist and believer in the American dream, this was a nightmare. The same talents that had made the country great were now threatening to lay it waste.

The unlikely Shakespeare of this dilemma, the political actor-dramatist most remarkably able to celebrate and lament the new order at the same time, was Ronald Reagan. To both supporters and opponents, Reagan seemed perfectly content to wield the nuclear threat as an instrument of overwhelming national power. In one notorious instance that briefly put Soviet military forces on alert, he joked into an open microphone, "My fellow Americans, I'm pleased to tell you today that I've signed legislation that will outlaw Russia forever. We begin bombing in five minutes."[18] Here was the essence of the modern nation-state in two sentences. Russia was reduced from a huge, diverse, complicated territory and its peoples to a legal abstraction that a competing state could simply repeal, with physical destruction following like an afterthought.

But Reagan grasped that the U.S. itself was exposed to the same danger. This was something he was often prone to fretting about. It wasn't just that America's own inventions had been turned against it, or that in doing so they had opened the U.S. homeland to easy attack for the first time in its splendidly isolated history. Reagan seemed to believe that something had gone wrong with the logic of history itself. Every offensive weapon had always generated a corresponding defense. Why had that stopped? How could a nation so powerful, so technically advanced, be completely defenseless apart from threatening the mass destruction of others?

There must be some solution—and at various times, Reagan and his administration proposed four: victory, abolition, space-alien invasion, and strategic defense. Early in his presidency, Defense Department planners floated proposals for a new doctrine of winnable nuclear war. At the Reykjavik summit in 1986, Reagan offered a ten-year plan for total nuclear disarmament. On several occasions, he spoke mistily of the common purpose that humankind might achieve if some mutual threat were at hand, like an attack from another planet. And his most enduring proposal was the Strategic Defense Initiative, an attempt—still unrealized—to neutralize the new offensive machines by inventing even newer defensive machines. Reagan spoke as if he wanted not just to end the nuclear threat but to vindicate technology itself. Because America is a technological nation, he seemed to think, this would vindicate America as well.

Although the search for strategic defense has outlasted the Cold War, on the whole Reagan was a quirky exception. No matter what a president might wish, a technical solution may not be available. (The alien-invasion scenario never panned out either, except in the movies.) The most successful ideologies work a different way: by making awkward questions simply unaskable. They embed a given set of arrangements so deeply into our ways of thinking and talking that they come to seem like reality itself—not the result of political choices, which can be made and unmade, but just the way things unavoidably are.

We have an excellent example of this from the Cold War in the work of Herman Kahn, one of the founders of the field of nuclear strategizing. Kahn was reportedly one of the inspirations for the maniacal title character in Stanley Kubrick's brilliant 1964 film satire, *Dr. Strangelove: Or, How I Learned to Stop Worrying and Love the Bomb*. Unlike Strangelove, though, who operated only in the (literal) shadows, Kahn to his credit went public, trying to explain for the benefit of ordinary citizens how theoreticians like himself had been training military and political leaders to think about the nuclear threat. In one "thought experiment," he asked his listeners to ponder the appropriate U.S. response if the Soviets were to level New York City in a one-shot nuclear attack. The strategically correct answer was to retaliate by nuking Leningrad—*not* Moscow, as many people wrongly assumed. Kahn would then go on to explain why.

But what about all the people this would kill? Kahn had thought of that too:

> The U.S. government might conceivably feel that the people of Leningrad were not implicated in the decision by their government to destroy New York City. Therefore, American leaders might want to punish the Soviet regime by destroying an important national treasure, while doing as little damage as possible to more or less innocent bystanders.
>
> Most people would argue, however, that citizens must suffer the consequences of acts taken by their government, whether they were explicitly consulted or not. One almost has to adopt such an attitude, even if it is close to the edge of immorality. There is almost no other rule that would work—at least in most such circumstances.[19]

It is obviously irrational to ignore "circumstances" or to do things that don't "work." The rhetorical construction of language here conceals, and thus tries to normalize, a certain construction of reality, in which the nation-state and its political goals have become simply "circumstances," which makes questioning them silly and naïve. (Dismantling such constructions, exposing the hidden meanings, was a project of *de*construction, another resource that Nuclear Criticism drew from literary analysis.) It is worth noting that the word *state* can also refer to "the state of things," the way things just are. By the Cold War, this double meaning had finally come into its own.[20]

But even Kahn recognized that he was "close to the edge," that his reflexive state-centrism was still competing with older, time-honored moral systems—as expressed, for instance, by the U.S. Conference of Catholic Bishops, which argued in a widely discussed 1983 pastoral letter that America's nuclear posture could not be reconciled with traditional church teachings on "just war." Nuclear threats were so extreme as to make them inherently unjust. Feeling compelled to reply, Kahn allowed that "spiritual and moral values certainly inform practical and secular ones, but"—he insisted—"they cannot substitute for them... It would be immoral and unwise to jeopardize our national security interests and the values of most of the world on the basis of a strongly held but emotional evaluation of basically technical and strategic issues."[21] Here the trick is to grant the legitimacy of moral concerns while categorizing and thus containing them. Safely tucked into slots labeled "spiritual" and "emotional," they can be isolated from the practical, technical and strategic. There may be some ultimate, metaphysical nature of things (the "spiritual"), and there are individual needs and urges (the "emotional"). But the world of states, or what Kahn calls "the real world," lies between these non-rational extremes. The real world is a world of technical issues best left to experts who know what works.

What Kahn was expressing is an *ideological* commitment to what works—the elevation of instrumental effectiveness from a servant of other values to a prime value in its own right. This commitment, it seems, is among "the values of most of the world," indeed a central value inasmuch as most of the world has been influenced by the West. It may be especially important in America, where there had always seemed to be a knack for getting things to work. Which, of course, is one way of honoring this value: by building something new that actually works, as Reagan hoped to do with strategic defense. But that way is risky, expensive and time-consuming. A much easier way is the one that Kahn demonstrates: just dismiss alternative values as sentimental and unreal. For a time, during the Cold War, both approaches succeeded, at least in the sense that events did not prove them wrong. Effective missile defense kept failing various tests, but further research at least allowed the dream to continue. And Kahn's thought experiments had the great virtue of never being tested at all. Had they been, it is no more likely that they would have gone the way Kahn imagined than it was likely that the Soviet Union would become the Communist utopia that Karl Marx imagined. The difference is that the Soviet collapse brought the latter dream to an end, while nuclear weapons and strategists are still very much with us. So are neoconservative and other war-planners more generally, and some of their experiments have recently escaped from thought into the actual world.

The results, as we know, have not been kind. What they have done, in fact, is vividly reveal the difference—the collision—between the ideology of what works and what actually works. This ideology, like any other, inflicts itself on reality and demands that reality adjust. Practical likelihoods are set aside, and conclusions are drawn on the basis not of observed facts but of a predetermined system of ideas. But the what-works theory is an anti-theory, a theory supposedly devoted to the purely practical. And so, ironically, like any other obsession, the purely practical winds up being pursued with wild impracticality. "What works" governs the policies of states *even when it doesn't work.*

Recent years have given us excellent illustrations of this in both fiction and real life. Consider two popular fictions in which—by contrast to *Dr. Strangelove* – the ideology of what works is celebrated rather than spoofed. Tom Clancy, who started writing thrillers during the Cold War, continued in the 1990s with a series of novels in which a CIA operative who just wants to do what works, "to get the job done," reluctantly succeeds to the presidency of the United States. As it happens, Iran has been expanding its influence in the Middle East, has taken over Iraq, and is now launching terror attacks within the U.S. So the new and inexperienced president finds himself waging a Middle East war against Iranian state-sponsored terror, in alliance with Saudi Arabia, using limited but high-tech American forces temporarily sited in the Saudi desert. This works; the job gets done.[22]

A few years after Clancy imagined these events, something like them actually happened, but in reverse: a once-temporary siting of forces in the Saudi desert became an ongoing U.S. presence; non-state-sponsored terrorists retaliated with attacks on the U.S.; and the response of an inexperienced president was a Middle East war, including a takeover of Iraq with a high-tech but "light footprint." This wound up increasing the influence of Iran. What Clancy foresaw as "the job's" motivating cause turned out, instead, to be its ironic *result.* And while these real events were unfolding, another action thriller, Joel Surnow's 2001-2010 television series *24,* followed the exploits of another American agent, a harried but resourceful terrorist-hunter who excuses his extreme tactics by explaining, "I had a job to do." Surnow carried Clancy's logic a step further. Where Clancy's way of foregrounding the "job" was to give us a president-operative who disavows any other political identity or goals, *24* radically pruned away the narrative altogether, leaving only the awful choices forced minute-to-minute on a hero who is barely allowed to pause for breath. Neither fiction invited us, or its characters, to ask difficult questions *about* the job: how it was defined, in whose interests, what unintended consequences it might have, and whether getting it done would be the end of our problems or just the beginning.

In another way, too, *24* went a step or two further. What put it on the cultural map was its sympathy for what actual U.S. operatives in those years were calling "enhanced interrogation." Like wars in the desert, like nuclear thought experiments, this practice was defended, in real life as on T.V., as the one that would "work" in the circumstances. But here again, the successes so evident in make-believe kept refusing to materialize in fact. As details and then photos of America's "War on Terror" leaked, the word *unthinkable* was brought out of Cold War storage—this time in reference not merely to theoretical events like nuclear exchanges, but to something that had already been tried, and judged, many times in the past. If nuclear threats were "close to the edge of immorality," as Herman Kahn had acknowledged, and if nuclear war might threaten civilization, torture was certainly beyond that edge; where it occurred, civilization had already been lost. That, at least, was the settled view of much of the world.

Those who defended torture's American renaissance therefore had to contend with a well-established body of laws, treaties, constitutional principles, judgments of courts and war-crimes tribunals, and a fairly recent international Convention Against Torture, promoted by Ronald Reagan himself, that forbade treating anyone, including non-citizens and official enemies, in ways that "shock the conscience." Getting around all these prohibitions required a lot of counter-arguing. In hundreds of pages of memos, with much mincing of words and chopping of logic, and with footnotes to dubious legal authorities as far back as Shakespeare's time, U.S. government attorneys struggled to prove that torture wasn't really torture, that it did not shock the conscience, but that if it did shock the conscience it did so in pursuit of the government's legitimate "interests," which therefore made it not shocking. As Herman Kahn had said of his other unthinkables, the only real question was effectiveness—and that, said the new enablers like John C. Yoo, was a technical question, for experts only:

> There can be little doubt that intelligence operations, such as the detention and interrogation of enemy combatants and leaders, are both necessary and proper for the effective conduct of a military campaign... Congress can no more interfere with the President's conduct of the interrogation of enemy combatants than it can dictate strategic or tactical decisions on the battlefield. Just as statutes that order the President to conduct warfare in a

certain manner or for specific goals would be unconstitutional, so too are laws that would prevent the President from gaining the intelligence he believes necessary to prevent attacks upon the United States.[23]

The verbiage was thick and the citations copious, but what it boiled down to was: the president and his agents have a job to do, and while they're getting the job done, the rest of us need to stay out of their way. There is no other basis for moral questioning. We became a "we" by yielding our individual consciences to the political goals of the nation-state, which aims to protect us. This is a "legitimate governmental objective," and as such it legitimates—in fact, makes "necessary"—the specific things done to that end. Conscience, therefore, has already spoken. From here on it is exercised on our collective behalf by the state, which in this case means the president, which in practice means operatives working in secret at prison camps and foreign "black sites."

The problem for torture apologists, as for nuclear strategists, is that there still are moral intuitions that make themselves felt apart from the state and without its authorization. On the basis of religious and other traditions, modern consciences have a stubborn way of continuing to be shocked. The efforts of modern nation-states to suppress that impulse have so far had only temporary and limited success. Much as they might want to push the envelope, to think outside the box, the authors of military and CIA doctrine still find themselves up against people who, literally and figuratively, didn't get the memo: bishops, peaceniks, opposition politicians, even colleagues in government who cling to quaint ideas, like the notion that there is a "rule of law" that limits even the state, or that some things are true and others aren't regardless of what the state and its leaders might wish. As one British official noted of his American counterparts, during secret discussions that laid the basis for invading Iraq, "the intelligence and facts were being fixed around the policy." Nonetheless, "the case was thin." It would soon be thickened (with the help of enhanced interrogation, no less).[24] But the problem with merely "fixing" the facts is that it concedes that facts matter. When these can no longer be denied—when it turns out, for instance, that Iraq had no weapons of mass destruction, or that its factions cannot agree on how the country should be run—the policy can still be discredited. Facts are a box; to think outside the box, you need to get beyond facts. Ideally, you would abolish the whole concept.

We might say that American policy in recent years made the mistake of relying too much on the old-school, Machiavellian solution. When he first glimpsed where modern politics was taking us, Machiavelli counseled princes that the key to effectiveness was hypocrisy: while their actual efforts should focus entirely on what is "useful" and gets results, which often means making war, they should still give the *appearance* of being faithful and upright, because people still naively believe in such things. Pretense is needed as long as truth and virtue can still be imagined. Maybe a further advance, though, will take us beyond them—from the Machiavellian solution to the Orwellian. Discrediting moral questions, rhetorically redefining them (like Herman Kahn) as "emotional" failures to recognize what "works," is a step in that direction. But it's an intermediate step, a temporary expedient. The final step, which Orwell foresaw at the start of the Cold War, is reinventing language altogether. "Newspeak," the state-crafted language of the sinister Inner Party in *Nineteen Eighty-Four*, would replace thought itself with propaganda, making questions contrary to the state's program impossible even to put into words. And in the meantime, for those not getting with the program there's enhanced interrogation. The Inner Party uses rats for this, the CIA used dogs and insects, but in each case the goal is to create a local reality in which the victim knows that every aspect of his experience is under the control of the state. And in *Nineteen Eighty-Four* the further aim is to leave him unable even to feel, let alone think, anything that isn't approved. Unending war frees the state from defeat, total surveillance roots out dissent, and thought control, finally, abolishes doubt—the ability to conceive of any other reality—even within the individual mind.

The Bush administration never got that far. But it had hopes:

> The [Bush] aide said that guys like me were "in what we call the reality-based community," which he defined as people who "believe that solutions emerge from your judicious study of discernible reality." I nodded and murmured something about enlightenment principles and empiricism. He cut me off. "That's not the way the world really works anymore," he continued. "We're an empire now, and when we act, we create our own reality. And while you're studying that reality—judiciously, as you will—we'll act again, creating other new realities, which you can study too, and that's how things will sort out. We're history's actors... and you, all of you, will be left to just study what we do."

That is how the president's chief political strategist explained things to a journalist in 2004, apparently unconcerned that talking like this made him sound like Josef Goebbels.[25] The journalist, Ron Suskind, said this remark went "to the very heart of the Bush presidency." It has been quoted often to that effect, and it did fit well with Bush's aggressive efforts to politicize a wide range of government functions. But it is probably best described as aspirational. We have, one might say, not yet arrived at nineteen eighty-four. Even for the most powerful managers of the world's most powerful state, "we create our own reality" is a hope, not a program. Stories, perhaps, can be made to order, but not history, at least not for real. But if this notion of "the way the world really works" does not, in fact, work, it expresses very well the *ideology* of what works. One ex-colleague dismissed the talkative official and his crew as so many "Mayberry Machiavellis." That suggests, though, that they knew they were dissembling. It would be more accurate to call them overmatched Orwellians, believers in their own visions who thought they could make them real. They just weren't very good at getting this to work.

Nostalgia

The nation-state's unresolved struggles with reality and morality help explain something else that I think we've seen highlighted in recent events: the way every war creates nostalgia for previous wars, including, as we might have predicted, the Cold War. This, too, is a phenomenon that calls for further attention from Nuclear and post-Nuclear Critics.

Even more than most human activities, war demands to be narrated. Because destruction, in itself, is intuitively shocking and wrong, it needs to be explained in order to make any sense. Explaining it means telling the story of what led to it or what it's supposed to achieve. In fact, it calls for stories within stories, a kind of narrative nesting-doll: this particular assault is meaningful only as part of this battle, which is meaningful as part of this war, which makes sense only in terms of the political aims of the warring parties, which derive from the epic mega-narratives whereby they account for their origins and destinies. Both the history and the literature of war, therefore, are full of grand stories, heroic tales of derring-do and noble sacrifice. In the classical odes, the individual warrior achieves glory and fame; in Shakespeare the humble but "happy few" are promised fame in odes yet to be written. Victory is helpful, but not essential: to one side, the American Civil War was a Battle Cry of Freedom, to the other it was a glorious Lost Cause. But either way it can be fashioned into appealing stories, even if the cause was actually dubious, the conduct savage, and the casualty rolls unimaginably long.

From early on, though, there is a persistent note of disillusionment, a suggestion that the narrative machinery might fail. Achilles, in the *Iliad*, threatens at one point to leave Troy unconquered and give up a life of renown: "Coward and brave man both get equal honour. Death treats idle and active men alike." Falstaff, Shakespeare's comic relief and an old friend of King Henry's, pauses from another great battle for an even more cynical riff: "Can honour set to a leg? no: or an arm? no: or take away the grief of a wound? no. Honour hath no skill in surgery, then? no. What is honour? a word." In the American Civil War, an ambulance driver rebukes a big shell that crashes near him: "You damn son of a bitch. You haint got no eyes, & would as soon hit a ambulance driver as anybody else." Sometimes the complaints are elegiac, as in the trench poetry of World War I, and sometimes satirical, as in antiwar black comedies like Joseph Heller's *Catch-22* (1961). Regardless, what dismays the eyewitness-critic is the sheer absurdity and arbitrariness of destruction, the fact that death is no respecter of persons and that all are equally vulnerable because all are made of flesh and blood.[26]

Technical innovations, new weapons and tactics, are difficult to fit into existing stories or the conventions for creating them. They will not be the weapons with which the old heroes won their famous victories. To the contrary, they will be alien to the old ways: probably more powerfully destructive, and designed to work by confounding the old, familiar warcraft. Also, very importantly, newer methods tend to scramble existing ideas of agency—shifting power, for instance, from individuals to massed groups, from human warriors to machines, or from daring skill to pushbutton automation. All of this makes them seem, if not diabolical, at best ignoble and amoral—mere intensifications of violence pure and simple. Sometimes they are resisted for that reason; anachronism is indulged and the obsolete kept on because of its link to past glories.[27] But given the advantage to be gained, however temporarily, from mastering new techniques before the other side seizes them first, such resistance is usually brief. Future wars incorporate the new methods and therefore tend to be fought differently than past wars. In modern times these changes have been greatly accelerated, first by the instrumentalist ideology of the nation-state and then by the industrial and information revolutions. Wars now differ significantly even when they're just a decade or two apart.

The ironic result is that each war re-emerges during the next as one of the "good old wars" that made the past a better place than the present. However atrocious it seemed at the time, the new atrocities of some future conflict will cast a haze of nostalgia over the earlier. There seem to be few if any exceptions to this. During the Civil War, observers on both sides recognized its unprecedented ferocity and worried that "God surely has forsaken his people. Given them over to destroy one another!"[28] Yet the innumerable stories told about that war ever since have continued to give us valiant volunteers, swashbuckling officers and flags flying in the thick of heroic charges. Likewise, at the time, Korea and Vietnam were dismal, inconclusive slogs against enemies who were hard to see; if anything they inspired nostalgia for the stirring tank battles and great naval engagements of the world wars. Yet during the Gulf War of 1991, even Vietnam was looked back on as "a real war," with beleaguered "grunts" in real firefights against formidable foes who didn't just surrender at the first chance—unlike this new conflict, which seemed more like a one-sided video game.[29]

I deliberately just skipped past the world wars. Those must have broken the pattern, right? Surely no one could be nostalgic for the mud and stupidity and industrial-scale killing of World War I? And wasn't World War II clearly a "good war" fought by "the greatest generation"? Well, at the time, not necessarily:

> War does horrible things to men, our own sons included. It demands the worst of a person and pays off in brutality and maladjustment. It has become so mechanical, inhuman, and crassly destructive that men lose all sense of personal responsibility for their actions. They fight without compassion, because that is the only way to fight a total war.

This was written about World War II just after it ended. The author, a veteran soldier, merchant seaman and war correspondent, went on—like a Joseph Heller who wasn't trying to be funny—to denounce the bureaucracy, regimentation, wastefulness and ineptitude of army life and its commanders: "It destroyed [the soldier's] personal sense of values, because a GI could not call even his soul his own."[30] And by contrast, at least in some minds, the earlier Great War had already passed into romantic myth; the term "Great" was subtly taking on a second meaning. Speaking in 1935, General Douglas MacArthur waxed as sentimental as any second-rate Victorian poet over "those days of old" that "have gone glimmering through the dreams of things that were. Their memory is a land where flowers of wondrous beauty and varied colors spring, watered by tears and coaxed and caressed into fuller bloom by the smiles of yesterday."[31] Smiles? On the Western Front? The First World War also gave us Sergeant York, who was not just a national celebrity, adventure hero and purported model for "G.I. Joe," but a modern example of Shakespeare's promise that war can elevate any humble provincial to greatness (even though Sergeant York's specialties, sniping and sharpshooting, were tactics that at other times have been seen as the *opposite* of heroic[32]). As another veteran, an old flying ace, reminisced, "You were treated like a soldier – they respected you." Looking back with satisfaction on a geopolitical disaster that so many, at the time, had scorned as stalemate and pointless carnage, the aging World War I flyboy offered this wistful homage, heard again and again when people look back: "That was a *real* war."³³

The Cold War had the disadvantage of not being a "real" war. Its long, twilight struggle was an affair of spies and shadows, punctuated by proxy fighting in various woebegotten locales and with the nuclear threat ever present. Yet it had not been over for long before it too became a good war, a model of "bipolar" clarity and ultimate success compared to the confusions that followed—the insurgencies and ethnic conflicts, the failed states and bafflingly complicated civil wars of the next two decades. And the innovations continue; the new infernal arts of today are cyber attacks and "drone wars":

> The mechanized, inhumane nature of these killings makes them even worse. With conventional war, there is at least a sense of honor, courage, and mutual respect among combatants. But with drones the victims are slaughtered like animals, with no chance to flee, to surrender, or to fight back – thereby at least gaining an honorable death. Nothing in the strikes acknowledges the humanity of those slain.
>
> [...] Soldiers in combat are themselves subject to risk. In a sense, a soldier is in a "kill or be killed" situation. Thus, when a soldier kills, the conscience, which strongly resists killing, is less injured. But a drone operator, remotely located, is not subject to risk or threat; his or her actions are mere killing (i.e., not motivated by a genuine instinct of self-preservation). Clearly this must have severe negative consequences for the psychological well-being of the operator. We are taking decent American young people and inducing them to be merciless killers and assassins.³⁴

Sneaking into another country's airspace or computer systems hardly seems gallant. Whatever happened to the good old days of, say, the Iraq occupation? Now that was a *real* war, with actual troops "in country," working "special ops" and "taking out hostiles" at sites with code names like "the Alamo," then lending a hand to rebuild schools and redirect sectarian grudges into soccer matches. Not like nowadays, when bloodless technicians stare at screens and push buttons from air-conditioned bunkers thousands of miles away.

Will the pattern continue, and the button-pushing also come to be looked back on fondly as "a real war"? It would not be surprising. The urge to redeem experience through inspiring stories seems as deeply planted in human nature as any other "Neanderthal mentality." But that enduring power of narrative is also grounds for cautious optimism. It has given us, over the millennia, continuing efforts to limit war and reduce its destructiveness. The notorious failings that lead to war—ambition, religious intolerance, greed, the imperatives of state—are closely related to the search for limits. Whatever people believe in strongly enough to fight over is also a source of values expected to endure beyond the fight. The fight must therefore be structured to keep those values intact. A warrior's ambition will be hard to "celebrate in song and story" unless his deeds exemplify some genteel code of honor. A war for religion will swallow the religion's own teachings unless rules are developed to keep it "just."[35] Greed is self-defeating unless benefits are weighed against longer-run costs. Even the instrumentalism of the modern techno-state has a possible upside in the attempt to professionalize war, to put it in the safe hands of sober theorists who have beaten their swords into spreadsheets.

The state, as Benedict Anderson argued, is itself a product of narrative—an expression of people's willingness to see themselves as fellow actors in a shared story. And one of its virtues is its flexibility, the fact that the particular stories attached to any given nation can change. This is a problem, and a *casus belli*, when some demagogue finds he can rouse people with a new (or new-old) story promising, say, to restore the lost glories of a Greater Serbia. But it's also a saving grace. Already in 1990, one prominent analyst correctly foresaw "Why We Will Soon Miss the Cold War"—but on some points he guessed wrong:

> Take away the present Soviet threat to Western Europe, send the American forces home, and relations among the [European Community] states will be fundamentally altered. Without a common Soviet threat or an American night watchman, Western European states will do what they did for centuries before the onset of the Cold War—look upon one another with abiding suspicion. Consequently, they will worry about imbalances in gains and about the loss of autonomy that results from cooperation. Cooperation in this new order will be more difficult than it was during the Cold War. Conflict will be more likely.

In this and other ways, the analyst feared, "a complete end to the Cold War would create more problems than it would solve."[36] There have certainly been problems, including tensions within Europe, especially in the wake of the 2008 financial crisis. But European states are not doing "what they did for centuries before," and I can't believe they will again. The Cold War didn't just distract them from their old aggressions, it redirected the old narratives. Europeans explain their experience to themselves and each other differently than they once did. Those states tell—in fact, are— different stories now, stories that include a "European project" for which the states and their citizens struggle in common. "Europe" is no longer just an empty stage on which German, French, and Russian stories play out and inevitably collide. And we also have new stories about what it means to be German, French, or Russian. Some old "characters" have been retired; the narratives no longer include German or French "races," let alone "the Aryan" or "the Hun" or "the Slav," prosecuting their alleged racial "destinies" from ethnically homogenous "homelands." For some purposes, at least, the *nation* part of "nation-state" has withered away: it is clearer today that virtually all states contain a diversity of nations and that no state wholly contains any one. Even in the chronically unstable Balkans, the lesson seems to be sinking in that visions of an ethnic Greater State are formulas for genocide and endless war.

We can reasonably hope that these developments have put war on a path of long-term decline, that it will join slavery as a once-common inhumanity that eventually becomes a synonym for "uncivilized."[37] No doubt there will still be lethal fighting, as there is today among "subnational actors" like gangs, militias and warlords. But this will be viewed as a crime, aberration or social problem—an absence of government authority, not an expression of it. The idea that war is "the continuation of politics by other means" will give way to an acceptance that other means are not available, that the only legitimate continuation of politics is more politics. The stages leading toward this endgame are already underway, each one marked by a new rule expanding the moral "duty of care":

1) Noncombatants on our own side should be protected if at all possible.
2) Noncombatants on the other side should be protected if at all possible.
3) Combatants on our own side should be protected if at all possible.
4) Combatants on the other side should be protected if at all possible.

Rule #1 arrived some centuries ago—but was nonetheless a moral awakening, not something that was always obvious. Rule #2 was largely accepted in the West by the mid-twentieth century, and Rule #3 by late in that century. We have seen #3 at work in the abandonment of military drafts and the switch to armies of volunteer professionals, in the "Powell Doctrine's" insistence on limited aims and "overwhelming force," in the strong preference for aerial strikes, commando raids, cruise missiles and now drones over "boots on the ground," and in the anxiety to keep troops out of harm's way even at the risk of failing (of allowing, for instance, the terrorist mastermind you've been chasing to escape through the mountains of Afghanistan).

If the logic holds, Rule #4 should follow. Like other rhetorical and political constructions, "the enemy" stands to be deconstructed—recognized as a temporary label loosely pasting together people with their own lives and realities. When that stage arrives, war becomes hard to imagine, let alone justify. Violence cannot be used to advance policy once policy comes to include protecting even adversaries from violence. Narratives that sustain war are also neutralized; they come to feel wrong, in the same way that old cowboy-and-Indian stories feel wrong when they just assume the villainy of Indians.[38]

Clearly, work toward this advanced ethic remains to be done. But there have been hints of it, auguries of a new consciousness even amid such erstwhile triumphs as the Gulf War. It did not escape some observers that there were moral problems with slaughtering Iraqi conscripts *en masse*—that what made them such easy targets for "turkey shoots" was precisely that they were, uniforms notwithstanding, mostly helpless victims of the dictatorship whose victims the war was meant to help. They had been conflated with their state and thereby cast in the narrative as the enemy, but not many years later, a different narrative cast them and others like them as a people (supposedly) awaiting liberation. Eventually, large numbers of Iraqi troops even wound up on the American payroll.

In a globalized world, we have to assume that any given group may thus be redefined, that those in our crosshairs today may need our protection tomorrow, that previously unlikely candidates may come to have an ethical claim on us—are, in fact, applicants for membership in our "we" and "us." It seemed strange to me, in the 1980s, that U.S. leaders extolled the Polish Solidarity movement while aiming nuclear warheads at the Gdansk shipyard, its home base. Not long after, the leader of Solidarity became president of a free Poland, which then joined NATO and the European Union. Today the shipyard's business partners range from a French yacht company to the government of Vietnam, it's a showpiece for EU-sponsored "green" efforts to replace nuclear power with wind turbines, and its Soviet-era "Lenin gate" was recently restored, partly to qualify the shipyard as an EU cultural heritage site and partly as a scenic backdrop for a movie about Solidarity. A onetime symbol of oppression still has value, but only for telling stories.

The future is unknown and therefore easy to worry about. Nostalgia redirects our hopes for it onto a safely idealized past; a story about the past is one we can better control, one that we know we have already survived. Like the other ways of thinking we've reviewed here that have helped to underwrite war, nostalgia is a means for making sense of experience by putting a satisfying construction on it, plotting it out with narratives and hierarchies and systems of value. If one role of the critic is to undo such constructions, that job is as important now as it was during the Cold War, or ever. Nation-states are still busy manufacturing reality to suit their political aims, and in today's world they work hand in glove with an aggressive global corporate and financial system, a so-called "free market" that presents itself not as a set of contestable policy choices but just as the way things must be. By exposing these pseudo-realities, criticism can help people navigate through them in search of a better way forward.

It might seem odd to read imaginative works in search of facts, or to look for reality in a future that hasn't happened yet instead of a past that already has. This, however, is the upshot of Martin Luther King Jr.'s famous belief that history is moving the right way, that the long "arc of the moral universe" bends toward justice. The Nuclear Critic may find inspiration in the source from which King borrowed that phrase: an 1852 sermon by Theodore Parker, one of nineteenth-century America's leading preachers and public speakers. Parker argued that justice is like gravity—a law of nature, one of the forces of attraction that hold the universe together. Most people know this; they are "always looking for the just," and "all this vast machinery which makes up a State, a world of States, is, on the part of the people, an attempt to organize justice." Parker saw a parallel here between the world of states and the world of literature, another arena to which people's search for justice takes them. Support for states is the political expression, and enthusiasm for fiction the "poetical" expression, of the same moral sense.

Of course, says Parker, there is still plenty of injustice. In the same way that physical laws allow for many "oscillations" and "aberrations," a world of billions of people necessarily has its pockets of immorality— what Parker calls temporary "bubbles." And of these, unfortunately, "the great national bubble which we call a State" is one of the most important. Injustice survives because leaders of states rely on realism, or so they think: "Not conscience and the right is appealed to, but prudence and the expedient for today. Justice is forgotten in looking at interest, and political morality neglected for political economy." Statesmen assume that morality "is no part of politics in the State." But this is *false* realism, at odds with "the facts of the world." Those facts—the forces actually operating, even in the affairs of states—are aligned with the humanities, not the inhumanities. "Tell me not of successful wrong," Parker says. "So far as anything is false, or wrong, it is weak." That's why the moral arc bends the right way. After that favorite phrase of King's, Parker's next words are these: "Things refuse to be mismanaged long." Because it just doesn't work.[39]

Notes

[1] Jonathan Schell, *The Fate of the Earth* (New York: Knopf, 1982). For a more detailed discussion of Schell and of antinuclear writing, see Jeff Smith, *Unthinking the Unthinkable: Nuclear Weapons and Western Culture* (Bloomington: Indiana University Press, 1989), esp. chs. 1-3.

[2] This notion, evidently a commonplace among management consultants, is brilliantly expanded on with reference to state action in Dwight Macdonald, "The Responsibility of Peoples" and "Massacre from the Air," reprinted in *Memoirs of a Revolutionist: Essays in Political Criticism* (New York: Farrar, Straus and Cudahy, 1957), 33ff and 72ff.

[3] Benedict Anderson, *Imagined Communities: Reflections on the Origin and Spread of Nationalism*, rev. ed. (London: Verso, 1991).

[4] Shakespeare, *Henry V*. The quotations are from IV.1 and IV.3. (I am omitting line numbers because these differ among editions, but searchable texts of all Shakespeare's plays are easily found online.) This analysis of *Henry V* is presented in more detail in Smith, *Unthinking the Unthinkable*, ch. 4.

[5] Again, many writers have described this development, but one I found particularly helpful was F.H. Hinsley, *Sovereignty*, 2nd ed. (Cambridge: Cambridge University Press, 1986).

[6] *Henry V*, IV.7.

[7] Michael Howard, *War in European History* (Oxford: Oxford University Press, 1976), 13.

[8] William Congreve, writing in 1806, quoted in Simon Werrett, "William Congreve's rational rockets," *Notes and Records of the Royal Society* 63, no. 1 (March 2009): 48, http://rsnr.royalsocietypublishing.org/content/63/1/35.full. Congreve was defending proposals for naval rocketry against claims that it lacked these "gallant" attributes. It is ironic, in light of the criticisms of earlier centuries, that in making this moral case for rockets Congreve relied on their *similarity* to old-fashioned artillery.

[9] Niccolò Machiavelli, *The Prince*, ch. 5, transl. Harvey C. Mansfield, 2nd ed. (Chicago: University of Chicago Press, 1998), 20-21.

[10] *Henry V*, III.3.

[11] Shakespeare, *Henry VI Part 1*, I.4. Although written before *Henry V*, the *Henry VI* trilogy (along with *Richard III*) deals with events that followed it—making *Henry V* an early example of what we now call a "prequel." We have no record of why Shakespeare wrote his various plays when he did, but one might speculate, in light of the observations below on nostalgia, that after telling the dismal tale of *Henry VI* he was looking to recall the glories of the earlier wars.

[12] *Henry VI Part 1*, IV.7, III.3.

[13] *Henry VI Part 1*, V.4, III.2, II.1 and III.1.

[14] James Davison Hunter, who popularized the term "culture wars" in its current usage, argued that "the long, tearful and sometimes bloody debate about Vietnam was not about the fate of a peasant society in southeast Asia, but about America. The debates about Lebanon and Panama and more recently the Persian Gulf were about America too. In each case, the opposing interests of the larger culture war

have been very much present and at play." Hunter, *Culture Wars: The Struggle to Define America* (New York: Basic Books, 1991), 289. For a more recent discussion linking Vietnam to the culture wars, see Craig Detweiler, *A Purple State of Mind: Finding Middle Ground in a Divided Culture* (Eugene, OR: Harvest House, 2008), esp. pp. 33-34.

[15] George Orwell, *Nineteen Eighty-Four* (New York: Harcourt, Brace, 1949). This unparalleled meditation on modern state-logic is discussed at greater length in Smith, *Unthinking the Unthinkable,* ch. 6.

[16] *Henry V*, II.4.

[17] The classic exposition of this recurring theme in American literature is Leo Marx's *The Machine in the Garden: Technology and the Pastoral Ideal in America* (Oxford: Oxford University Press, 1964, 2000). Marx's analysis and its application to Reagan and SDI are further detailed in Smith, *Unthinking the Unthinkable,* ch. 5.

[18] Andrew Glass, "Reagan jokes about bombing Russia, August 11, 1984," *Politico*, Aug. 11, 2010, http://www.politico.com/news/stories/0810/40921.html.

[19] Herman Kahn, *Thinking About the Unthinkable in the 1980s* (New York: Simon and Schuster, 1984), 63.

[20] Of course, the nuclear state did not invent this convenient definition of "circumstance." President Andrew Jackson, addressing American Indians during their "Trail of Tears" expulsion from the southeastern states, wrote: "Circumstances that cannot be controlled, and which are beyond the reach of human laws, render it impossible that you can flourish in the midst of a civilized community." Those "circumstances," he neglected to add, were certainly in part of his own making. Jackson, "To the Cherokee Tribe of Indians East of the Mississippi River," March 16, 1835, *Teach US History*, http://www.teachushistory.org/indian-removal/resources/cherokee-tribe-indians-jackson.

[21] Kahn, *Thinking About the Unthinkable in the 1980s*, 216.

[22] Tom Clancy, *Executive Orders* (New York: Berkeley Books, 1997). See also Jeff Smith, *The Presidents We Imagine: Two Centuries of White House Fictions on the Stage, on the Page, Onscreen, and Online* (Madison, WI: University of Wisconsin Press, 2009), esp. 206, 227-230.

[23] John C. Yoo, "Re: Military Interrogation of Alien Unlawful Combatants Held Outside the United States," U.S. Department of Justice Office of Legal Counsel memorandum, March 14, 2003, *ProPublica*, http://www.propublica.org/special/missing-memos?s=1. For further discussion and background on the OLC memos, see David C. Cole, *The Torture Memos: Rationalizing the Unthinkable* (New York: New Press, 2009), 1-40. Other memos particularly relevant to my discussion (and also web-posted at *ProPublica*) include Stephen G. Bradbury's, May 30, 2005, and Jay S. Bybee, "Interrogation of al Qaeda Operative," August 1, 2002, a document strikingly similar to the kind of memo that one imagines Orwell's Inner Party would have generated on Winston Smith, the hero-victim of *Nineteen Eighty-Four*—including references to personal phobias of the detainee's that can best be used against him.

[24] David Manning, "Iraq: Prime Minister's Meeting 23 July," July 23, 2002, *The Downing Street Memo(s),* http://downingstreetmemo.com/memos.html. The British preferred an ultimatum, something Iraq would refuse so it would then become, in the words of Shakespeare's Henry, "guilty in defence." But, the memo reported, "It seemed clear that Bush had made up his mind to take military action," and "There was little discussion in Washington of the aftermath after military action."

[25] As the columnist Molly Ivins once quipped—referring to Pat Buchanan, another former American minister of propaganda—it probably sounds better in the original German. The Bush aide, reputedly Karl Rove, is quoted in Ron Suskind, "Without a Doubt," *New York Times Magazine,* Oct. 17, 2004, http://www.ronsuskind.com/articles/000106.html. The term "Machiavellian" was also applied to this kind of talk; see, for instance, Lawrence M. Ludlow, "Machiavelli and U.S. Politics," August 15, 2005, *Future of Freedom Foundation,* http://www.fff.org/comment/com0508d.asp. But, as noted below, I think that usage misses a helpful distinction. For a further discussion see Smith, *The Presidents We Imagine,* esp. ch. 7.

[26] The quotations are from the *Iliad,* Book 9; *Henry IV Part 1,* V.1; and J.E. Lendon, *Soldiers and Ghosts: A History of Battle in Classical Antiquity* (New Haven, CT: Yale University Press, 2006), 47 (quoting the memoirs of Confederate Gen. Edward Porter Alexander). The key scene for present purposes in *Catch-22* is the death of the radio gunner, Snowden, in chapter 41.

[27] See Lendon's discussion in *Soldiers and Ghosts,* esp. ch. 2, of the Greek taste for anachronisms intended to assimilate real war to the imagined past. A modern example is the use of antique swords, horse-mounted troops and the like in the "parade" and "color" displays and uniforms of contemporary armies.

[28] See the quotations in Harry S. Stout, *Upon the Altar of the Nation: A Moral History of the Civil War* (New York: Viking, 2006), 211-212 and *passim.*

[29] This analogy, common in news commentary, was famously expanded to book length in Jean Baudrillard's *The Gulf War Did Not Take Place* (1991; Bloomington, IN: Indiana University Press, 1995). For other examples, references and critiques, see Matthew G. Kirschenbaum, "Part V: Radical Artifice," *A White Paper on Information,* University of Virginia, 1998, http://www2.iath.virginia/edu/mgk3k/white/frames.html, and Melanie Swalwell, "'This isn't a computer game you know!': revisiting the computer games / televised war analogy," *Digital Games Research Association*, November 2003, http://www.digra.org/dl/db/05150.18371.pdf. In an interesting twist, a noted battle from the Gulf War later became an actual video game, the forerunner of many digital simulations now used in U.S. military training. See "This is not a video game," *The Economist*, March 3, 2012, http://www.economist.com/node/21548490.

[30] Edgar L. Jones, "One War is Enough," *The Atlantic Monthly* 177, no. 2 (February 1946): 48-53, http://www.theatlantic.com/past/docs/unbound/bookauth/battle/jones.htm. For other examples of criticisms like these, see Macdonald (note 2 above) and Paul Fussell, "The Real War 1939-1945," *The Atlantic Monthly* 264, no. 2 (August 1989): 32-48, http://www.theatlantic.com/past/docs/unbound/bookauth/battle/fussell.htm.

[31] Gen. Douglas MacArthur, "The Noblest Development of Mankind" (speech to the 42nd Infantry "Rainbow" Division reunion), July 14, 1935, *American Experience*, http://www.pbs.org/wgbh/amex/macarthur/filmmore/reference/primary/macspeech01.html.

[32] For instance, early in Euripides' *Heracles*, two characters debate whether the legendary hero (also known as Hercules) is smart or cowardly for leaving the line of fire, keeping under cover and shooting at enemies who can't see him—in other words, for doing with a bow exactly what Sergeant York does with a rifle.

[33] "Van Hampton Burgin – WWI Flying Ace," *Voices from the Past: A Burgin Genealogy Website*, http://theburginfamily.org/familymatters/matters04.html. A Google search of the phrase "now that was a real war" in early 2012 produced references (in that same sense) to the following, just within the first ten "hits": Korea, Vietnam, the American Revolution, the Crimean War, the Māori resistance to the British in New Zealand, and of course World War II.

[34] John Uebersax, "On the Immorality of Drone Missile Strikes in Pakistan," letter to Sen. Barbara Boxer, March 25, 2011 (misdated 2012), web-posted as "Drones1" at http://www.john-uebersax.com/plato/.

[35] For an illuminating side-by-side comparison of war-limiting doctrines in both Christianity and Islam, see James Turner Johnson, *The Holy War Idea in Western and Islamic Traditions* (University Park, PA: The Pennsylvania State University Press, 1997), esp. ch. 5.

[36] John J. Mearsheimer, "Why We Will Soon Miss the Cold War," *The Atlantic Monthly* 266, no. 2 (August 1990): 35-50, http://teachingamericanhistory.org/library/index.asp?document=713.

[37] I would like to say that torture is another such practice, but its recent and, one hopes, temporary American revival is a reminder that progress of the sort I'm describing is fitful and precarious, often a matter of two steps forward and one step back.

[38] Robert Wright and Stephen Pinker, among others, have predicted similar developments in their recent articles and books, although from sociological rather than "literary" viewpoints. It would make an interesting study to catalog the narrative genres and formulas that have lost their salience as a result of what we once called "consciousness-raising." Besides cowboys and Indians, popular fictions of the last hundred years routinely trafficked in race and gender ideas that now look hopelessly anachronistic, they assumed the virtues of colonial domination and white supremacy (or "Aryan," as it was called, even in America), and they made heroes of people doing things we now deplore, like whaling and big-game hunting. Because most such stories are discarded once their ideological assumptions become visible, it is easy to forget how common they were even recently, and easy to overestimate the staying power of current genres, including war stories.

[39] Theodore Parker, "Of Justice and the Conscience," in Frances Power Cobbe, ed., *The Collected Works of Theodore Parker, Vol. II: Sermons, Prayers* (London: Trübner, 1879), 37-57. In closing, I would like to thank Alice Shira Phillips

Walden, Deborah Kearney, Bob and Margye Smith, and the editors of this volume for ideas, encouragement, and helpful guidance in the writing of this essay.

Chapter Two

Specters of Totality: The Afterlife of the Nuclear Age

Aaron Rosenberg

The "institution of Nuclear Criticism," proposed at a Cornell University conference and in a subsequent *diacritics* special number in 1984, tasked itself with providing a forum for professionals in the humanities to intervene meaningfully in the nuclear issue. The conference became something of a mini-movement, attracting Jacques Derrida as its keynote speaker and drawing on a timely interest in deconstruction. This type of Nuclear Criticism was, to be sure, critical theory of the literary variety—yet its political and ethical commitments were no less serious for it. Indeed, a critical intervention was considered urgent at a time when apocalyptic rhetoric concerning an immanent total nuclear war was framing political, scientific, and strategic decisions, but was not being assessed in terms of how certain kinds of nuclear discourse could themselves produce very real threats.

Nuclear Criticism responded by calling for a "rhetorical analysis of the forms, the themes, the performance of nuclear political argument as it is presently enacted," an analysis that would ultimately question "all the forms of nuclear discourse which obey rhetorical constraints, which submit to forms of censorship, exploit narrative figures and tropological devices in order to persuade."[1] These efforts would serve to deconstruct the language of nuclear authority, along with claims to "expertise" that conjoined technical, scientific, political, and military knowledge. Of the nuclear situation Derrida insists, "We are certain that, in this area in particular, there is a multiplicity of dissociated, heterogeneous competencies. Such knowledge is neither coherent nor totalizable...the frontier is more undecideable than ever."[2] Therefore, the rigorous application of *textual* criticism and interpretation was considered essential to developing an understanding of the cultural logic of the "nuclear age," a period in which the specter of the absolute end—a fable of the future—

was shaping the reality of the present.

For the Nuclear Critics, what distinguished the nuclear age as historically unique from all other periods was not simply the destructive force of its weapons, but moreover their unprecedented capacity to weaponize language itself. In his seminal formulation of this argument, "No Apocalypse, Not Now (full speed ahead, seven missiles, seven missives)," Derrida maintains that although "there is perhaps no invention, no radically new predicate in the situation known as 'the nuclear age,'" nevertheless, "nuclear weaponry depends, more than any weaponry in the past, it seems, upon structures of information and communication, structures of language, including non-vocalizable language, structures of codes and graphic decoding."[3] Derrida recognizes in nuclear proliferation a syncretizing process whereby language and technology become increasingly co-determined. By relying on structures of communication and information exchange, nuclear weapons systems expose themselves to the same aporias that belong to language practices in general. Further, the runaway escalation of global nuclear arsenals pressurizes the communicative structures upon which these systems ultimately depend, such that the speed of decision-making during a nuclear crisis tends to become split-second, without time for discussion, debate or negotiation—meaning that exchanges must be *predicted* in advance for all kinds of contingencies.[4] And of course it is not difficult to recognize how all forms of communication can go awry: at an interpersonal level, communication is vulnerable to misunderstandings, misinterpretations, and misinformation; among nations, even the most seemingly transparent statements can be distorted by cultural differences, rhetorical posturing, and diplomacy; and formal logic, the language used to interface with technology, and technology with itself, is subject to syntactical problems, malfunctions, and human error.[5] This state of affairs, as the nuclear critics noted, becomes ironic—with global nuclear arsenals wired into error-prone linguistic systems, total annihilation might be triggered not by choice, but by accident.[6]

In "No Apocalypse, Not Now," Derrida further suggests that in the case of total nuclear war these problems are inflected by the mode of fiction. This is because total nuclear war names an event that has never as yet happened in history, an event about which there cannot be any direct experience. As Derrida points out, this hypothetical, total event—the absolute limit to which nuclear discourse ultimately refers—cannot be considered "real"; it must instead be understood as a literary, virtual "phenomenon whose essential feature is that of being *fabulously textual*, through and through." To the extent that the total phenomenon can be said to "exist," it only exists in language, "one can only talk and write about it."[7] Rhetoric that engages with this essentially virtual event points only to a "signified referent."[8] Thus, to discuss, speculate or strategize about total nuclear war is to engage with a fiction, a text—to perform a kind of literary criticism.

However, Derrida's insistence on the textual character of the nuclear situation does not imply that its material effects are any less real. The fable becomes real through a process of *mediation*, whereby the virtual event produces its own "reality," what Derrida grandly calls the "encompassing institution of the nuclear age."[9] What is "constructed by the fable" is nothing short of:

> This fabulous war effort, this senseless capitalization of sophisticated weaponry, this speed race in search of speed, this crazy precipitation which, through techno-science, through all the techno-scientific inventiveness that it motivates, structures not only the army, diplomacy, politics, but the whole of the human *socius* today, everything that is named by the old words culture, civilization, *Bildung, scholè, paideia*."[10]

In effect, Derrida's essay asserts that in the nuclear age, human society itself can potentially be read as the reification of a nuclear fiction, "as if," as Richard Klein explains, "[l]iterature were the motor of capital formation."[11]

I offer this rehearsal to underscore some of the ways in which mid-1980s Nuclear Criticism envisioned the nuclear age broadly, as a period indexically related to literary and linguistic crises. By reinterpreting the limits, both the beginnings and endings of a history called "the nuclear age," the Nuclear Critics opened a space for serious engagement with the strange metonymies, cause-and-effect relationships, and forms of anticipation that condition the relationship between the literary and "the real." What occupied both the center *and* the limit of the critique was total nuclear war, the technological capacity to erase all human life. Despite—or perhaps because of—the enormity of this concern, few academics seem willing to continue the project of Nuclear Criticism today. Total nuclear war, along with its totalizing language, is no longer timely. Klein has recently acknowledged that the "real or major event" about which he and Derrida were writing "belongs to the nuclear age, which most have thought came to an end when the Cold War ended."[12] Total nuclear war is no longer discussed as a reality of the *foreseeable* future even though its materials (both fissile and textual) remain stubbornly present, and it is unclear how the rhetoric of the mid-1980s can or should be updated in order to address current nuclear concerns. We are faced with different anticipatory anxieties than those discussed at the Cornell conference—the nuclear events of today's future are imagined as sudden, but not total, or else total, but no longer sudden. In 1994, Christopher Norris could already claim that "what ha[d] changed" in the ten years since the arrival of Nuclear Criticism was "that highly specific conjuncture—of rhetorical 'escalation' to the point of aporia of absolute 'undecideability'—from which this movement first took rise and in which it discovered a short-lived pretext for some fairly arcane and wire-drawn argumentation."[13] Norris also points out that Nuclear Criticism's commitment to negotiating the figurative language of an all-encompassing, total nuclear war, the fable that occasioned its important questions, paradoxically subjected it to the same all-or-nothing discourse it had sought to critique. And, as others have argued, attempting to align the distinct events that form the basis of real experiences within a nuclear history—including the actual bombings of Hiroshima and Nagasaki, as well as the many nuclear tests and accidents that have killed real people—to a totalizing, "fabulously textual" event that would constitute the endpoint of such a history, can be radically reductive.[14] Yet the critique that most destabilizes Nuclear Criticism, that its preoccupation with total nuclear war makes it inadequate for dealing with specific nuclear events, often overlooks how nuclear events, whether real or imaginary, resist specificity by spilling over geographic, temporal, or other "imagined" borders—whose figures of waste and decay, radiation and fallout, are cruelly non-specific. Nevertheless, it would seem that in order for Nuclear Criticism to speak to the situation of the present, whose proximally-anticipated events are of a partial, limited, or local nature, it would need to abandon rhetoric involving "the inescapable catastrophe, the undeviating precipitation towards a remainderless cataclysm," that today seems hyperbolic, even nostalgic.[15]

However, to dismiss the problematic concept of total nuclear war because it lacks immanence would be premature. Nuclear weapons continue to exist with the force to destroy life as we know it (many times over). But there are other ways in which the *figure* of total nuclear war, which was essential to the deconstructive strategies of Nuclear Criticism, can continue to offer a means of reading the linguistic systems and literary tropes that would structure a new fable of humanity's post-Cold War nuclear survival. By this I am suggesting that the absence of apocalyptic rhetoric from today's nuclear discourse may perhaps indicate the presence of a new kind of fable, one that surmounts, represses, or defers the eschatological anxieties of the former. In this way, discussing the nuclear age as a past historical period can be read as an attempt to assert closure, to consign to it an epithet of technological obsolescence (as with the "space age")—thereby arrogating to ourselves the "wherewithal to neutralize invention, to translate the unknown into a known, to metaphorize, allegorize, domesticate the terror."[16] In short, we risk considering ourselves competent with the total nuclear threat simply by virtue of living in a "post-apocalyptic" time.

Thus, even though for the present we can no longer say, as Derrida did, that we are living *in suspense*—that "the worldwide organization of the human *socius* today hangs by the thread of nuclear rhetoric"—I would suggest that we must be attentive to what happens when the referent of an absolute end becomes suspended.[17] How can the absence of a sudden, totalizing crisis be reconciled with the lingering knowledge that the threat remains a possibility, however distant? And how must the temporality of the nuclear age be reconsidered, now that the figure of total nuclear war has become *doubly* fabulous, belonging as it does to an alternative past that did not happen, and still, perhaps, to a speculative future to come?

***

In thinking about what is at stake over claims that the nuclear age has ended, one must first ask when it can be said to have begun. Maurice Blanchot, in his 1964 essay "The Apocalypse is Disappointing," identifies the start of the nuclear age along these lines: as the era that commences with the prospect of total annihilation *and* as one in which this prospect produces a new vision of global community. He begins by paraphrasing Karl Jasper's dramatic proclamation, "Today there is the atomic bomb; humanity can destroy itself; this destruction would be radical; the possibility of a radical destruction of humanity by humanity inaugurates a beginning in history, whatever precautionary measures there may be, we cannot go backward."[18] Blanchot argues that while Jaspers is content to claim that this beginning in history has irreversibly changed the conditions of human existence, he does not demonstrate any real commitment to changing radically his philosophical priorities. Further, Blanchot cautions that we must be vigilant to the ways in which the danger of total nuclear war can be instrumentalized to entrench the same political positions that produced its conditions of possibility. Since the nuclear situation is phrased in terms both urgent and final, as one in which "either man will disappear or he will transform himself," it becomes clear that "this transformation will not only be of an institutional or social order; rather, what is required in the change is the totality of existence."[19] Blanchot argues that calls for change on this scale must be assessed in terms of their commitment to wholly different ways of thinking. Otherwise, the nuclear situation "only serves as an alibi or a means of pressure for bringing us to spiritual or political decisions that have already been formulated long ago and independently of it."[20] Here is a kind of ideological opportunism in which "the event, the pivot of history, does not change the options or the fundamental oppositions in the least...reflection on the atomic terror is a pretense; what one is looking for is not a new way of thinking, but a way to consolidate old predicaments."[21]

Whatever form of collective "humanity" is made possible by the condition of the nuclear age, it cannot be said to antedate the prospect of the total nuclear event: "What does the problematic event teach us? This: that insofar as it puts into question the human species in its totality, it is also *because of this event* that the idea of totality arises visibly and for the first time on our horizon—a sun, though we know not whether it is rising or setting; also, that this totality is in our possession, but only as a negative power."[22] This claim demonstrates how the prospect of total nuclear war carries an organizing force commensurate to its destructive potential: to imagine the erasure of all human life is to gain a vision of collective humanity, menaced by a universal peril. However, the awareness of the threat, potently figured by Blanchot's metaphor of the sun, does not tell us whether the nuclear age signals the dawn or the twilight of such a vision; the radical negativity of the total nuclear event yields an unstable and incomplete kind of knowledge:

> Through understanding, we know very precisely what must be done in order for the final annihilation to occur, but we do not know what resources to solicit to prevent it from occurring. What understanding gives us is the knowledge of catastrophe, and what it predicts, foresees, and grasps, by means of decisive anticipation, is the possibility of the end. Thus man is held to the whole first of all by the force of understanding, and understanding is held to the whole by negation. Whence the insecurity of all knowledge—of knowledge that bears on the whole.[23]

In discussing the danger faced by humanity as a whole, it is fair to ask at this point if the Nuclear Criticism of the 1980s was concerned with purely nuclear issues, or whether its real commitments were to analyzing what Blanchot goes on to call an "avatar of totality"—one that emerges within, but is perhaps not limited to, the nuclear age.[24] Yet it is clear that for Nuclear Criticism totality was not just historically topical, but also essential for understanding the precarity of language that points to an ultimate destruction without trace. Lest we discount these broad concerns as either exaggerations of the manifesto form, or else as "merely academic" thought experiments, we need only remember that Derrida could write with absolute sincerity that, in 1984, nuclear weapons had "put in the balance the fate of what is still now and then called humanity as a *whole*, or even of the earth as a whole."[25] By responding with a project for rethinking the nuclear situation in terms of its language, the Nuclear Critics were engaging with Blanchot's main concern: "Why does a question so serious—since it holds the future of humanity in its sway—a question such that to answer it would suppose a radically new thinking, why does it not renew the language that conveys it…?"[26]

Derrida's version of nuclear historicity follows the logic of Blanchot's account—that the radical negativity made possible by the total event inaugurates the nuclear age—but when this understanding is paired with an insistence on the event's textuality, it produces a quite unexpected departure. While the possibility of total nuclear war first occurs in the twentieth century and makes visible, simultaneously, the beginning and the ending of a certain collective vision of humanity, Derrida proposes that this condition had already been prefigured by the condition of "literature," which "comes to life and can only experience its own precariousness, its death menace and its essential finitude. The movement of its inscription is the very possibility of its effacement."[27] Derrida proceeds to argue that the "fabulously textual" nuclear event, itself a fiction, emerges from within the same literary history that it ultimately threatens to erase. And it is in this critical move that we hear the strongest echoes of Blanchot:

> If "literature" is the name we give to the body of texts whose existence, possibility, and significance are the most radically threatened, for the first and last time, by the nuclear catastrophe, that definition allows our thought to grasp the essence of literature, its radical precariousness and the radical form of its historicity; but at the same time, literature gives us to think the totality of that which, like literature and henceforth in it, is exposed to the same threat, constituted by the same structure of historical fictionality, producing and then harboring its own referent. We may henceforth assert that the historicity of literature is contemporaneous through and through, or rather structurally indissociable, from something like a nuclear epoch.[28]

It is important to note Derrida's formulation of the structural relationship between literature and the nuclear age as "indissociable," rather than analogical. Because the nuclear situation is expressed as a fiction, it can be said to arise from the condition of literature even as its emergence threatens literature's "body of texts." Thus, the absolute end menaces all of humanity not only in the sense of aggregate human life—of species-being—but also in the totality that constitutes its "symbolic capacity." What gives "humanity" as a whole a sense of meaning, according to

Derrida, is its "movement of survival...the '*survivance*,' at the very heart of life."[29] This "movement of survival" is indissociable from what Derrida has earlier called the "movement of inscription"—the possibility for creating meaning over time by writing and recording, of adding to the body of texts that perform "monumentalization, achivization, *work on the remainder, work of the remainder.*"[30] The total nuclear event forecloses the very possibility of meaning, since the instantaneous and final end erases the time necessary for "movement," for the work of mourning and *survivance*. This radical negativity in turn exposes the fragility, the tenuousness of the "juridico-literary archive," which is not simply a repository of texts; rather, it names the corpus of textual relationships that bestow meaning to language, no less than the quasi-material "collection" that organizes memory into encompassing collectivities—such as history, culture, and society.

Ultimately, Derrida seems to be suggesting that the nuclear age, like and within literature, can only have one end, toward which it is always pointing—the radical effacement of its conditions of possibility, perhaps the final cross-over from the fabulously textual to "the real." Defined in this way, the historical present of the nuclear age is not organized by the presence of a total event at all, but rather by its suspension or deferral. This helps to make sense of Derrida's claim that "the nuclear age is not an epoch, it is the absolute épochè; it is not absolute knowledge and the end of history, it is the épochè of absolute knowledge."[31] In referring to the nuclear age as the absolute épochè, Derrida invokes Husserl's sense of épochè—as a bracketing and deferral, postponing the "judgment before the absolute decision."[32] And perhaps we may discover in this figure of the absolute épochè a way of understanding the nuclear age, not as a period that ends with the cease-fire of global superpowers, but as an ongoing present in which the final act has been, and is being, continually suspended.

***

The legacy of the nuclear age, then, may be the endurance of this avatar of totality: the vision of humanity arranged under the sign of its absolute, irreversible end—the end motivated by its own inventiveness. And, if so, the anticipated total nuclear event belonging to the Cold War may represent only the first iteration of an enduring change to the conditions of human existence. Indeed, the kind of collective citizenship underwritten by nuclear annihilation could be considered a precursor to globalization—still fertile ground for criticism. Today's anticipatory critical models continue to make use of discourse involving irreversible endings—particularly, for example, ecocriticism. Molly Wallace has recently suggested that "bringing Nuclear Criticism and ecocriticism together under the rubric of something like a 'risk criticism'...might offer a way to theorize the mega-hazards of the present."[33] One begins to recognize that an avatar of totality inhabits the discourse of global risk paradigms in general, along with claims that humanity as a whole must radically change its thinking in order to survive. The force of absolute destruction, which galvanizes the very idea of a shared humanity, should also organize its collective struggle against its death drive. In an era in which we are all potential victims, we are also charged with imagining the means of averting our terrible, common fate. But this, too, constitutes part of the nuclear fable; it presupposes that the pronoun "we" names an extant collective agent capable of addressing the risk that circumscribes it. Blanchot, for one, contests this assumption: "It is not even true that the radical destruction of humanity is possible," he argues, "for it to be possible, one would need the conditions of possibility to be united: real freedom, the achievement of the human community, reason as a principle of unity... a totality that must be called—in the full sense—communist."[34] Notwithstanding the political alternatives capable of supporting collective agency, it becomes clear that any common effort "we"—as underwritten by shared exposure—might make to change the precarious status quo would inevitably involve rewriting its organizing fictions.

Accordingly, it is necessary to question whether arguments for change that invoke totalizing discourse trace the persistence of the nuclear age or a departure from it. In 1984, the same year as the Nuclear Criticism *diacritics* issue, Lyotard delivered this famous protest: "We have paid a high enough price for the nostalgia of the whole and the one... we can hear the mutterings of the desire for a return to terror, for the realization of the fantasy to seize reality. The answer is: Let us wage a war on totality."[35] Yet even as he aims to subvert the totalizing conditions of his nuclear age present, Lyotard nevertheless reproduces its rhetoric in calling for a war on such a scale. And just as Fukuyama's "end of history," as Derrida argues in *Specters of Marx*, named precisely what it wasn't, the 1991 collapse of the Soviet Union—the putative end of the nuclear age—did not result in the discontinuation of totalizing political rhetoric. One finds specters of nuclear age totality in the post-9/11 War on Terror, the kind of war "waged *in the name of...*" that Derrida alludes to in 1984, one that anticipatorily frames diverse conflicts as a unified, global clash of ideologies.[36] The agglomeration of various "evil" security threats into a "whole" acts not only as a means of broadening interventionist foreign policy doctrine, but also as a language practice that reasserts superpower geopolitics by naming a spectral (and potentially permanent) antagonist.[37] Such events could be read broadly among accounts that submit a fissioning of the nuclear age—a model in which the nuclear age ends with the Cold War, but its fallout becomes scattered across the past, present, and future. But even this model, which could help to explain the next chapter in nuclear fable after the Cold War, cannot entirely dispel the abiding fiction of total nuclear war as humanity's final end, so long as its real conditions of possibility are still present.

While the rhetoric of totality endures, discourse specific to total nuclear war seems to have been subdued (albeit temporarily) by a consensus that it has become "a thing of the past." It is possible to recognize within this consensus an unconscious effort to rewrite the nuclear age in order to re-contain the total event. Defining the nuclear age as coextensive with the end of the Cold War could therefore be interpreted as a way to provide closure for certain kinds of anticipatory nuclear traumas. In "Bombing and the Symptom: Traumatic Earliness and the Nuclear Uncanny," Paul K. Saint-Amour, a contributor to this volume, considers whether "certain traumatic responses to the use of nuclear weapons might not have been limited to the survivors of Hiroshima and Nagasaki, but rather shared by all who knew of nuclear weapons, their devastating effects, and the escalating likelihood of their use."[38] Saint-Amour identifies in particular what he calls "the *hysteron proteron* of the nuclear condition: the literally preposterous phenomenon of traumatic symptoms…that exist not in the wake of a past event, but in the shadow of a future one."[39] If anticipatory anxieties can indeed produce this form of response, then the future possibility of total nuclear war will continue to traumatize. An historical narrative that positions events of the Cold War, such as the Cuban Missile Crisis, within a series of near-misses or false alarms may even serve to reinforce the symptoms. As Saint-Amour explains in "Air War Prophecy and Interwar Modernism,"

> Unlike the realized physical violence of a raid, a false alarm provides no catharsis for the sense of endangerment it produces; it mobilizes anxiety without providing it with a kinetic outlet. Thus the very falsity of the alarm emphasizes a condition of hideously prolonged expectation, a state of emergency that is both perennial, in having been detached from the arrival of violence in a singular event, and horribly deferred—the advance symptom of a disaster still to come."[40]

This deferral to the future involves a process of archivization, a kind of burial that can only be partial so long as predictions that are still possible remain unfulfilled and open. Thus, a collective fiction that provides closure for the nuclear age by relegating it to the past could be inscribed in the archive as a means of hastening what Derrida calls "the symbolic work of mourning, with memory, compensation, internalization, idealization, displacement"—the movements of survival.[41] But in the case of total nuclear war this symbolic mourning could only amount to an incomplete feeling of having survived what was always-already a non-event.

The singular, textual status of total nuclear war continues to pose temporal paradoxes even if it is assigned a past tense—even if it is now "experienced" as an event that *did not* happen, rather than one that *has not yet* happened. As Saint-Amour emphasizes,

> Anticipation alone, we should note, cannot traumatize. The repressed can only appear to return from the future, can only signal some looming eventuality in symbiosis with some past repression or wounding; only the thing that has already, in some sense, happened can be the cause of a traumatizing anticipation that imagines the disaster returning to complete its work. For these reasons, we misconstrue the apocalyptic imagination if we understand it as referring only to the future. For the prospect of that future revelation to wound, there must have been not only a prior concealment but also a prior revelation—prior yet incomplete, one whose consummation has been deferred, conditionally, to the future.[42]

If trauma originates from some form of prior experience, how can it be initiated by an event that, by definition, cannot be meaningfully experienced since it negates the possibility of a future? Derrida offers a possibility: "The event has never happened (except in fantasy, and that is not nothing at all)."[43] His claim relies on Freud's assertion that "there [is] no difference in the unconscious mind between reality and a fiction loaded with affect."[44] Accordingly, the only point when a total nuclear war could ever be said to "have happened" (that is, have been experienced) would be in the unconscious, which under certain conditions cannot distinguish fantasy from reality. As Klein explains, "In our unconscious, the worst may have already happened, and we are compelled to repeat the trauma in our dreams."[45] Taken together, these statements suggest that although an actual total nuclear event could only be atemporal, its occurrence in the unconscious—experienced as real—can assume temporal dimensions. But again we are returned to a contradiction of priority, since the total event, even in fantasy, should leave no remainder—no trace. This phenomenon corresponds to what Derrida elsewhere discusses as "the very experience of aporia":

As its name indicates, an experience is a traversal, something that traverses and travels toward a destination for which it finds the appropriate passage. The experience finds its way, its passage, it is possible. And in this sense it is impossible to have a full experience of aporia, that is, of something that does not allow passage. An aporia is a non-road.[46]

While the nuclear fable, as a fiction, mobilizes human activity toward its end, the end itself is aporetic, "nonnarratable." D.A. Miller distinguishes the "nonnarratable elements of a text" as "precisely those that…serve to supply the specified narrative lack, or to answer the specified narrative question. It is not the case that such elements cannot be mentioned. The nonnarratable is not unspeakable. What defines a nonnarratable element is its incapacity to generate a story. Properly or intrinsically, it has no narrative future."[47] This is why the closure of the nuclear fable, which ends in total nuclear war, can only be partial. "In essence," Miller writes, "closure is an act of 'make-believe,' a postulation that closure is possible," and while there may appear to be "moral advantages of this postulation" it remains marked by "its self-betraying inadequacy."[48] Telling ourselves that the nuclear age is over may help lay to rest the ghosts of the Cold War, but it cannot yet offer us the assurance that the threat of total nuclear war is gone. In this, we can only hope to repress our lingering fears, even as we anticipate the return of the end.

Notes

[1] Richard Klein, "Proposal for a *diacritics* Colloquium on Nuclear Criticism," *diacritics* 14, no. 2 (1984): 3.

[2] Jacques Derrida, Catherine Porter, and Philip Lewis, "No Apocalypse, Not Now (full speed ahead, seven missiles, seven missives)," *diacritics* 14, no. 2 (1984): 22.

[3] Ibid., 21, 23.

[4] Derrida emphasizes how this sense of speed arises from a linguistic and technological equivalence in his parenthetical subtitle, "full speed ahead, seven missiles, seven missives."

[5] The combination of these produce devastating consequences in Stanley Kubrick's *Dr. Strangelove* (1964): interpersonal communication is systematically derailed by Gen. Jack Ripper; the Soviets delay the crucial announcement of their ultimate deterrent, the doomsday device, for internal political reasons; and the all-important code for recalling the bombers (itself an acronym of "peace on earth" or "purity of essence"), once decrypted, fails to reach the last B-52 because of a radio malfunction.

[6] See Richard Klein and William B. Warner, "Nuclear Coincidence and the Korean Airline Disaster," *diacritics* 16, no. 1 (1986): 2-21; and Derrick De Kerckhove, "On Nuclear Communication," *diacritics* 14, no. 2 (1984): 71-81.

[7] Derrida, "No Apocalypse," 23.
[8] Ibid.
[9] Ibid.
[10] Ibid.
[11] Richard Klein, "Knowledge of the Future: Future Fables," *diacritics* 38, no. 2 (2008): 174.
[12] Ibid., 173.
[13] Christopher Norris, "'Nuclear Criticism' Ten Years On," *Prose Studies* 17, no. 2 (1994): 136.
[14] K.K. Ruthven argues that Derrida opens himself to his own "aporia" by discussing the bombing of Hiroshima and Nagasaki as events marking the end of a conventional, rather than a nuclear war. Ruthven maintains that one cannot translate these real events into the "fabulously textual," and that not all conflicts involving or overshadowed by nuclear weapons can be read as anticipations of total war. K.K. Ruthven, *Nuclear Criticism* (Carlton, Vic: Melbourne UP, 1993), 73.
[15] Derrida, "No Apocalypse," 21.
[16] Ibid.
[17] Ibid., 25.
[18] Maurice Blanchot and Elizabeth Rottenberg, *Friendship* (Stanford: Stanford UP, 1997), 101.
[19] Ibid.
[20] Ibid., 103.
[21] Ibid., 104.
[22] Ibid., 105 (emphasis added).
[23] Ibid.
[24] Ibid., 108.
[25] Derrida, "No Apocalypse," 22.
[26] Blanchot, 103.
[27] Derrida, "No Apocalypse," 27.
[28] Ibid.
[29] Ibid., 28.
[30] Ibid.
[31] Ibid., 27.
[32] Ibid.
[33] Molly Wallace, "Will the Apocalypse Have Been Now? Literary Criticism in an Age of Global Risk," in *Criticism, Crisis, and Contemporary Narrative: Textual Horizons in an Age of Global Risk*, ed. Paul Crosthwaite (New York: Routledge, 2011), 16.
[34] Blanchot, 107.
[35] Jean-François Lyotard, *The Postmodern Condition: A Report on Knowledge* (Minneapolis: Minnesota UP, 1984), 82.
[36] Derrida, "No Apocalypse," 30.
[37] The rhetoric of evil, though it doubtless appears in conflicts throughout history, also specifically marks the continuity of nuclear discourse within and after the

Cold War. In the early 1980s, President Reagan famously dubbed the Soviet Union "the evil empire." President George W. Bush echoed the phrase in his national address on 9/11, remarking, "Today our nation saw evil." Months later in his State of the Union Address, Bush conflated the evil of terror with the presumptive nuclear capabilities of Iran, Iraq and North Korea, branding those regimes the "axis of evil."

[38] Paul K. Saint-Amour, "Bombing and the Symptom: Traumatic Earliness and the Nuclear Uncanny," *diacritics* 30, no. 4 (2000), 61.

[39] Ibid.

[40] Paul K. Saint-Amour, "Air War Prophecy and Interwar Modernism," *Comparative Literature Studies* 42, no. 2 (2005), 140.

[41] Derrida, "No Apocalypse," 28.

[42] Saint-Amour, "Air War Prophecy and Interwar Modernism," 155.

[43] Derrida, "No Apocalypse," 23.

[44] Ibid.

[45] Klein, "Knowledge," 174.

[46] Jacques Derrida, "Force of Law: The 'Mystical Foundation of Authority,'" trans. Mary Quaintance, *Cardozo Law Review* 11 (1990), 947.

[47] D.A. Miller, *Narrative and its Discontents: Problems of Closure in the Traditional Novel* (Princeton: Princeton UP, 1981), 4-5.

[48] Ibid., 267.

Chapter Three

Queer Temporalities of the Nuclear Condition

Paul K. Saint-Amour[1]

When Jonathan Schell's *The Fate of the Earth* first appeared in 1982, its most talked-about passage was a graphic description of what would happen if a twenty-megaton bomb were detonated over the center of Manhattan. The ensuing account of how a full-scale nuclear change would likely extinguish humankind along with the majority of earth's species, leaving a "republic of insects and grass," completed the book's infernal vision. Largely owing to this vivid thought-experiment, Schell's book helped reenergize the anti-nuclear movement in the U.S., and its cautionary portrait of a dead, irradiated planet was absorbed into mass-culture such that, read now, it chastens but does not stun. But there *is* a still-astonishing moment in *The Fate of the Earth*. This occurs in a section called "The Second Death," where Schell adopts "the view of our children and grandchildren, and of all the future generations of mankind, stretching ahead of us in time." A nuclear extinction event, he argues, would wipe out not only the living but all of the unborn as well; this "second death" would be the death of a longitudinal, progenerative human future, the death of the supersession of generations and thus, as he puts it, "the death of death."[2] That we live in the shadow of the death of death, says Schell, is nowhere more apparent than in our growing ambivalence toward—and here is the surprise—*marriage*, an institution that consecrates a personal relationship by connecting it to the biological continuity of the species. "[By] swearing their love in public," he writes, "the lovers also let it be known that their union will be a fit one for bringing children into the world." In a world overshadowed by extinction, the biological future that endows love with social meaning begins to dematerialize, and love becomes, in response, "an ever more solitary affair: impersonal, detached, pornographic. It means something that we call both pornography and nuclear destruction 'obscene.'" Although Schell is not explicit about what forms of sexual detachment he laments here, "The Second Death" clearly implies that any sex decoupled from biological continuity and seeking refuge in licentious, solitary, distant, or momentary enjoyment—any sex that deviates from a reproductive notion of the future—is a symptom of our nuclear extinction syndrome. Thus when Schell, oddly quoting Auden, says that the peril of extinction thwarts "Eros, builder of cities," he doesn't need to invoke "sodomy, destroyer of cities" for a link between queerness and extinction to be forged.[3] By installing a reproductive futurism at the heart of his admonitory project, Schell implicitly stigmatizes as futureless anyone who stands beyond reproductivism's pale: not just the homosexual but also the unmarried, the divorced, the impotent, the childless, the masturbator, the hedonist, the celibate.

Schell's book did not, of course, invent the use of reproduction as a metonym for human futurity *tout court* or the figuration of the biological child as the chief beneficiary of future-oriented actions in the present. But it contributed to these figures' prominent standing in the anti-nuclear imaginary. "Believe me when I say to you / I hope the Russians love their children too" went the absurd refrain of Sting's 1985 single, "Russians," which placed the (implicitly reproductive) body at a level more fundamental than political difference: "We share the same biology / Regardless of ideology." One could go on to compile a long list of 1980s movies, novels, speeches, and tracts that made the nuclear family stand in for humanity's beset future or invoked the child as the figure in whose name apocalypse must be averted or at least survived. These conventions would outlast the Cold War and the waning or reimagining of the nuclear referent. Think of P. D. James's 1992 thriller *The Children of Men*, whose protagonist must safeguard a miraculous pregnancy in a future where fertility has declined globally to zero.[4] Or of how Cormac McCarthy's *The Road* (2006) pares the matter of survival in a post-apocalyptic, ambiguously nuked landscape down to a father's efforts to protect his son from rape and cannibalism. In both cases, the future is hanging either literally or allegorically by the thread of a single imperiled child.

My aim in this essay is not to trace the reproductivist energies of Cold War anti-nuclear works or of more recent post-apocalyptic fiction. Instead, I chart an alternate path through the nuclear condition,[5] one that diverges from—and in places dissents from—the portrait of a future secured primarily for the sake of the biological child and reached along the straight lines of reproductive heterosexual coupling, familial property heritage, and linear time. This alternate path is one on which Nuclear Criticism today might keep company with recent work on queer temporalities, a body of scholarship that places dissident sexuality in a critical relation to normative models of time and history. One of my broader aims, in fact, is to indicate some of the ways Nuclear Criticism might be reenergized by an encounter with queer temporalities scholarship. At the same time, I'll argue that some of the key theoretical and literary works associated with Nuclear Criticism in its early years were themselves engaged in queering temporality and history. In doing so I don't wish to claim Nuclear Criticism as the occulted or lost "origin" of queer temporalities work; in addition to straining credibility, such a privileging of origin in a narrative of linear development would install queer temporalities scholarship in just the sort of historical narrative it seeks to vex by its devotion to non-linear modes—the recursive, the discontinuous, the counterfactual. My point is, rather, that reexamining Nuclear Criticism through the aperture of queer theoretical writings on time allows us to see a muted or latent critique in the former—a critique whose object was not so much the existence of nuclear weapons as the straitened portraits of desire, culture, kinship, history, and futurity that were often appealed to in calling for both those weapons' abolition and their necessity. What emerges is a redrawn Nuclear Criticism that both deplores the existence of nuclear weapons and declines to embrace sexually normative and historically reductive grounds for their elimination.

"Queer temporalities" as a theoretical rubric covers a broad range of scholarship by queer theorists and activists working, at least to date, predominantly in the U.S.[6] More specific than a turn toward time as theme, this scholarship considers how heternormative cultures perceive queer subjects in relation to history and futurity; how queer subjects experience and enact particular relations to history and futurity; and how queerness itself might be rethought as having less (or less exclusively) to do with sex and sexual typology than with dissident ways of being in relation to time. I have already referred to one of the chief temporalities from which queer subjects are variously excluded and dissenting: the "reproductive futurism" that conscripts the child as mascot for a heternormative politics of hope and a linear conception of history as both powered and figured by biological reproduction and the modes of inheritance and political succession it undergirds.[7] Such a conception of history militates against certain kinds of transgenerational affect, not least against the notion that the living could invest affectively in or form communities with the dead. In response, some scholars working on queer temporalities advocate just such a queer desire for history or "touch of the queer," the kind of unpunctual, affective approach that could permit one to ask, as Carolyn Dinshaw does, "How does it feel to be an anachronism?"[8] While acknowledging that the feeling of being out of step with one's contemporaries can be exploited to repressive ends, Dinshaw remains optimistic that transtemporal communities—living anachronisms in league with the dead—might produce politically salutary effects in a present whose dense multiplicity they help to restore.[9] Others, contrastingly, refuse a politics of hope they see as irreducibly heternormative, urging queer subjects to embrace the negative position assigned them by reproductivism. Embracing this negativity can take many forms: an insistence on the destructive, anti-communitarian, at once selfish and self-shattering dimensions of sex and particularly homo-sex; an identification of the queer subject with destrudo (i.e., the Freudian death drive) in its relentless opposition to a procreative understanding of libido; or a refusal of queer triumphalism and an embrace of the shame-laced backward look.[10] Still others look to fuse the negativity of these anti-social, arguably apolitical positions to a radical anti-racist and anti-capitalist stance, calling for a "punk negativity" whose oppositional politics declines the language of hope, redemption, and futurity and turns instead to vandalism, masochism, pessimism, and despair.[11] Real differences inhere among these approaches. But they share a core conviction: that temporality—and perhaps futurity even more intensively than historicity—cannot be thought apart from the sexual norms through which it is figured, licensed, and imbued with or emptied of affect.

Owing to its semi-dormancy since the early 1990s, Nuclear Criticism has largely missed the chance to think through queer theory, a field whose principal interventions have happened in the interim. You occasionally see comparisons between queer coming-out narratives and a nation's coming out as a nuclear power or a military person's coming out as an anti-nuclear activist. But the more suggestive commonalities between Nuclear Criticism and queer theoretical writing—most of them under the sign of temporality—remain unexplored. These include an intimate acquaintance with and even an embrace of the death drive; a related acquaintance with portraits of the future as negated or foreclosed; a commitment not to reopen the future under repressive terms; and the alternative, in the face of a seemingly barred future, of soliciting the queer touch of the dead whom for various reasons we suddenly apprehend as our contemporaries. Exploring these commonalities seems the more urgent, given that queer temporalities scholarship could provoke debate about what nuclear abolitionists and their opponents have most in common: a practically automated recourse to reproductive futurism in arguing for their respective positions. Schell's equation of low marital indices with a general sense of species futurelessness is an extreme but not an exceptional case of anti-nuclear rhetoric, which continues today to invoke "a world safe for our children" in terms nearly indistinguishable from the pro-nuclear side of the aisle.[12] The radical negativity exhibited by some queer temporalities scholars might also expose the limits of a politics of (procreative) optimism on both sides of the nuclear debate—the limits of acting as if the world could be made safe for "our children" or anyone else by either retaining or abolishing our nuclear deterrents.

Queer theorists, for their part, have turned occasionally during the last twenty years to Nuclear Criticism, although usually to jump-start an argument headed away from the nuclear referent. Peter Coviello's essay "Apocalypse from Now On" (2000) nods in its title to both Jacques Derrida's inaugural work of Nuclear Criticism, "No Apocalypse, Not Now (full speed ahead, seven missiles, seven missives)" (1984) and Susan Sontag's 1989 *AIDS and Its Metaphors* (Sontag: "Apocalypse is now a long-running serial: not 'Apocalypse Now' but 'Apocalypse from Now On'").[13] But Coviello's essay invokes the nuclear condition principally in order to set up what he sees as its succession, after 1989, by AIDS as the apocalypse du jour. "Du jour" in the way a daily special marks the everyday's domestication of the exceptional: for Coviello, AIDS differs from the nuclear condition in quotidienizing apocalypse, making it a condition rather than a threatened event and thus particularly useful to the day-to-day biopolitical operations of the state. Coviello, in other words, sets sail from Port Derrida for Port Sontag—from Nuclear Criticism to a critique of AIDS and governmentality—without, understandably enough, booking return passage. Before leaving the nuclear behind, however, he notes "how intimately bonded the nuclear and the sexual actually were, before the advent of AIDS gave to such bonding a ghastly quality of inevitability."[14] Coviello's emphasis is not on the usual string of references to the heteronormative sexualization of nuclear weapons (e.g., "Little Boy," Bikini atoll, the population bomb, and the nuclear family, although he mentions these in passing). Instead, he reads nuclear discourse as having limned, before AIDS, a "gay death drive" that figured queerness as incarnating (and more rarely as rebuking) the extravagant sovereignty of nuclear weapons. Glossing Martin Amis's characterization of the nuclear arsenal as a cocked gun in the mouths of the procreative, Coviello writes that "*power in the nuclear age is horrifying and unlivable because it makes me—or wants to make me—thoroughly, irremediably queer.*"[15] Thus the homophobia of certain anti-nuclear discourses anticipated homophobic responses to AIDS as an apocalyptic threat emanating from queer subjects. What's more, Coviello hazards, the apocalypticism that pervaded debates around both nuclear weapons and AIDS made for strong continuities between Nuclear Criticism and queer theory, both bodies of work responding to high concentrations of state power in the management of populations, bilateral depictions of the biological family as under siege, and the pervasive rhetoric of the death wish.[16]

We have already seen how hospitable both pro- and anti-nuclear writing could be, and for that matter remains, to reproductive futurism and a range of homophobic rhetorics. One of the signal features of Nuclear *Criticism* (as opposed to anti-nuclear discourse writ large) has been to read nuclear discourses of all kinds with skepticism and thus to hesitate on the threshold of a reflexive apocalypticism. One result of this circumspection, I suggest, is that Nuclear Criticism is a repository of alternatives to heteronormative portraits of the future. Derrida's "No Apocalypse, Not Now" serves as my cardinal example. I do not revisit it for the warrant of professional competence it gives humanists writing on the nuclear debate or for its bold claim that literature has always belonged to the nuclear epoch, both of which I have discussed elsewhere.[17] My interest is in two other of the essay's features: its unconventional portrait of the futurity imperiled by nuclear war and its critique of the use of the name as chief rationale for both making and avoiding war. Two years after *The Fate of the Earth*, "No Apocalypse, Not Now" refers to the extinction hypothesis advanced by Schell and others—"that in nuclear war 'humanity' runs the risk of its self-destruction, with nothing left over, no remainder"—as a "rumor," adding that in the speed race of the nuclear condition, a few seconds "may decide, irreversibly, the fate of what is still now and then called humanity—plus the fate of a few other species."[18] The sardonic end to that sentence registers an impatience with anti-nuclear rhetoric's shrill anthropocentrism and anticipates the critique of the human/animal binary that emerges in Derrida's work of the 1990s. What's more, although the essay goes on to reclaim rumor as the discursive ground of the nuclear condition ("one can no longer oppose belief and science, *doxa* and *episteme*, once one has reached the decisive place of the nuclear age"),[19] it absolutely declines to join Schell et al. in invoking humanity, human progeny, or human biological survival as the thing in whose name Nuclear Criticism ought to proceed. In the course of responding to its other, more assertively phrased provocations, commentators on the essay have largely missed its stunning refusal to take up this most ready-to-hand rhetoric of the anti-nuclear movement.

Where "No Apocalpyse, Not Now" invokes the question of survival, it does so not on behalf of biological individuals, genealogies, or species but in respect to self-referential, self-constituting elements of the symbolic order that could not be regenerated out of lived experience or by way of scientific method. These hothouse flowers include written as opposed to oral literary modes and literary criticism—anything that would perish without the archive from which it derives its conditions of possibility.

> Now what allows us perhaps to think the uniqueness of nuclear war, its being-for-the-first-time-and-perhaps-for-the-last-time, its absolute inventiveness ... is obviously the possibility of an irreversible destruction, leaving no traces, of the juridico-literary archive—that is, total destruction of the basis of literature and criticism. Not necessarily the destruction of humanity, of the human habitat, nor even of other discourses (arts or sciences), nor even indeed of poetry or the epic; these latter might reconstitute their living process and their archive, at least to the extent that the structure of that archive (that of a nonliterary memory) implies, structurally, reference to a real referent external to the archive itself.[20]

What kinds of writing, Derrida asks, could be regenerated in the wake of the archive's destruction by a nuclear war? Certain literary modes (lyric, epic) and discourses (sciences, non-literary arts) whose archive is grounded in a "real referent" external to itself might return; not so, forms of literature whose chief referent is the "stockpile" of textual antecedents and legal precedents that give literature its internal meaning and its social significance. The latter forms, Derrida adds, have brooded on their precarity since their emergence in the Enlightenment—so much so that their total effaceability is also their transcendental referent. "The only referent that is absolutely real is thus of the scope or dimension of an absolute nuclear catastrophe that would irreversibly destroy the entire archive and all symbolic capacity, would destroy 'the movement of survival,' what I call '*survivance*,' at the very heart of life."[21] Texts pondering the erasibility that is the precondition of writing; a criticism dedicated to preserving the archive of its own possibility and thus the *survivance* at the very heart of life: against the heteronormativity of reproductive futurism, the queerness of an *archival* futurism.

We might object that a call to preserve the archive must be made, at least implicitly, *in the name of* something other than the archive itself—in the name of some future user or of the dead whose traces repose in and constitute the archive. "No Apocalypse, Not Now" closes with a meditation on names and naming that responds, at least in part, to such an objection. A war that destroyed all symbolic capacity would negate the transmissibility of any name in which it was waged. As Derrida puts it, and with a repetitive cadence that mimics invocation, "that name in the name of which war would take place would be the name of nothing, it would be pure name, the 'naked name.' That war would be the first and the last war in the name of the name, with only the non-name of 'name'...

Beyond all genealogy, a nameless war in the name of the name."[22] No worldly entity—not nation, not dynasty, not ideology, not genealogy—could justify a war that threatened to annihilate all worldly entities. Yet curiously, having cited genealogy as the thing transcended by the "naked name," Derrida forbears to call on genealogy as the rationale for *not* waging a nuclear war. He abstains, too, from invoking a name ("our children," "posterity," "the future of humanity") for whose benefit the archive should *not* be destroyed. Instead, the essay offers an alternate vision of a divine entente with the sons of Shem, who built the tower of Babel "to make a name for [them]selves." Here, instead of escalating to the point where the tribes are scattered and thrown into linguistic confusion, the war between God and the sons of Shem is suspended out of mutual ignorance that "they were confronting each other in the name of the name, and of nothing else, thus of nothing. That is why they stopped and moved on to a long compromise." The seeming fullness of the name is what produces "deterrence" and with it the long monolingual standoff in which "tradition, translation, transference have had a long respite."[23]

But nuclear war is the absolute knowledge, Derrida continues, that God and the sons of Shem lacked and that we possess—the knowledge that a war that would put an end to all naming could only be waged in the name of the name itself, and of nothing else, thus of nothing:

> We have absolute knowledge and we run the risk, precisely because of that, of not stopping. Unless it is the other way around: God and the sons of Shem having understood that a name wasn't worth it—and this would be absolute knowledge—they preferred to spend a little more time together, the time of a long colloquy with warriors in love with life, busy writing in all languages in order to make the conversation last, even if they didn't understand each other too well.[24]

What follows that "Unless," suspending the lethal name game that Derrida anatomizes, is the essay's vision of survival in the shadow of nuclear war. It is not a vision in whose name, or of those in whose name, war should be waged or averted. Knowing the irreversible limit toward which the logic of the name in war tends, it sidesteps that logic in favor of tarrying and writing—in favor of the writing that is a form of tarrying, the "tradition, translation, transference" that allow the conversation to last in all languages despite or because of imperfect comprehension. The monolingual name, with its power to license war *or* disarmament for the sake of a particular territorialized future, was the weapon that had to be laid down all along.

This is not the place to survey the importance of Derrida's work in general for queer theory.[25] But even in the foregoing brief discussion of archival futurism in "No Apocalypse, Not Now," we can see certain strong compatibilities with queer temporalities scholarship: in its quiet demurrer from reproductivism and species egotism; in its commitment, amid precarity, to refusing security on the wrong terms; and in its view of the archive as the primary site of long colloquies across time. In the hands of queer theorists, something like this survival or *survivance* of the archive indispensably grounds a conception of queer history as a "touch across time." Rather than knot Nuclear Criticism more deeply into recent queer temporalities scholarship, however, I will devote this essay's second half to a transtemporal reach in the other direction, reading "No Apocalypse, Not Now" by the light of an earlier Cold War post-apocalyptic speculative fiction – indeed, reading it implicitly *as* a work of this type of fiction. Here I mean to point up something other than the essay's much-discussed speculation as to Nuclear Criticism's coming institutionalization in the university.[26] My focus will instead be Derrida's hypothetical *mise-en-scène*: a devastated future in which the human race survives but confronts a "total and remainderless destruction of the archive."[27]

Imagine for a moment a post-apocalyptic novel endowed with the reproductive futurism of Schell's "Republic of Insects and Grass" but in which *Homo sapiens* is *not* extinguished by a nuclear holocaust. A story of endurance and resourcefulness against the odds, it tells how a few hundred survivors return from the brink of extinction, struggle to produce viable offspring, and labor to reestablish kinship structures in the face of a hostile, irradiated environment. Key elements of this multi-generational narrative involve questions of tribal structure, hierarchy, and scarce resources as well as the tradeoffs between endogamy and exogamy, monogamy and polygamy, primogeniture and other forms of birthright and property dissemination. As the genetic mutations catalyzed by high radiation levels begin to appear, the survivors' colony depicted by the novel must also find ways of ensuring that the mutants don't reproduce and create more disabled, possibly sterile offspring who will weaken the collective chances of survival. For those chances depend on strong, healthy, fertile offspring who will help guarantee the future of humanity—and whose faith in that future will be indexed by their willingness to marry and procreate. Subplots might address the loss, rediscovery, or transmission of pre-apocalyptic knowledge and technologies. But because this imaginary novel defines survival exclusively as biological perdurance, it will be interested in questions of literacy, technology, and the archive only insofar as they abet an unbroken species perpetuation.

Now imagine the inverse: a post-apocalyptic novel that foregrounds the loss of literacy and scientific knowledge as well as the problems of preserving the written remains of the pre-collapse civilization; a novel of archival futurism in which the survival that matters has only minimally to do with questions of parentage, inheritable property, clan, courtship, and reproduction. We have just such a novel in Walter M. Miller, Jr.'s *A Canticle for Leibowitz*, an early Cold War classic that appeared between 1955 and 1957 in the *Magazine of Fantasy and Science Fiction* and was republished, with alterations, in 1960. Miller's book has been studied for its engagements with Catholic theology and witness, its transgenerational ethics, its cyclical model of history, and a host of other traits. But we're now equipped to read *A Canticle* as something else besides: an encyclopedia of non-reproductivist forms of narrative, character, historiography, and anti-nuclear politics. The novel opens 600 years after a global nuclear war, the Flame Deluge of the late twentieth century, and is set in the desert of what used to be New Mexico. In an abbey there, the monks of the Albertian Order of the Blessed Leibowitz gather, guard, copy, and memorize the Memorabilia, the paltry written remains of the pre-collapse civilization. These writings have survived both the Deluge and a kind of second Babel or "confusion of tongues" culminating in the Simplification—the destruction, by the survivors, of the knowledge-archive they blame for having made the war technologically possible. The surviving documents include everything from circuit diagrams and physics textbooks to even more abstruse messages ("*Pound pastrami, can kraut, six bagels—bring home for Emma*"; "*Remember—pick up Form 1040, Uncle Revenue*").[28] Written in pre-Deluge English, the fragmentary Memorabilia are largely unintelligible to the monks, who lack the scientific knowledge, linguistic competency, and range of contexts needed to restore their sense. Yet they protect the documents nonetheless.

> It mattered not at all to them that the knowledge they saved was useless, that much of it was not really knowledge now, was as inscrutable to the monks in some instances as it would be to an illiterate wild-boy from the hills; this knowledge was empty of content, its subject matter long since gone. Still, such knowledge had a symbolic structure that was peculiar to itself, and at least the symbol-interplay could be observed. To observe the way a knowledge-system is knit together is to learn at least a minimum knowledge-of-knowledge, until someday—someday or some century—an Integrator would come, and things would be fitted together again. So time mattered not at all.[29]

Although the promise of an Integrator-to-come lends a messianic structure to the monks' vigil, the emphasis here is less on a particular future in whose name the Memorabilia should be preserved than on the intrinsic value of keeping intact a singular and internally consistent symbolic system, regardless of its intelligibility. Without hope of benefiting themselves, without even the motivation of a scriptural promise that the Memorabilia will one day become legible, the members of the Order maintain their vigil over the millennia.

Which means that *A Canticle for Leibowitz* also spans millennia. Where the Schellian novel I imagined above might have encompassed several generations—enough to get its protagonists through war, social and environmental collapse, and genetic mayhem to the beginnings of a new societal and procreative equilibrium—Miller's book is massively longitudinal, its three books being separated by 600-year intervals.[30] As if to underscore its dedication to divergent models of continuity and futurity, its protagonists are celibates whose parentage goes unremarked, whose childhoods are largely dark to us, and whose values, vows, offices, and property are passed on to other celibates through apostolic succession. Because *A Canticle* traces the *Bildung*, or re-*Bildung*, of literacy and technology across 1800 years from the Flame Deluge, it cannot tarry over the development of individual characters. The novel's commitment to the *longue durée* of civilizational recovery shapes its conception of protagonism, which is distributed not only among a range of characters during each of its three books but also across those books. Characters who occupy us in the earlier books are absorbed into the remote historical backgrounds of subsequent ones, their deaths narrated with pitiless brevity or simply taken for granted. Instead of playing to the temporality of the individual, the novel is keyed to a different "organism, the community whose cells were men, whose life had flowed through seventy generations… The organism lived as a body, worshipped and worked as a body, and at times seemed dimly conscious as a mind that infused its members and whispered to itself and to Another in the lingua prima, baby tongue of the species."[31] One effect of this passage is to resignify the word "species," which refers less to *Homo sapiens* than to the collective organism of the monastic community, a kind of eukaryotic algal colony that perpetuates itself through the recruitment of celibate human cells.

Miller's novel has frequent recourse to such metaphors of biological form and descent, as well as to figures of (usually patriarchal) kinship. Monks are addressed as "Brother" or "Son," nuns as "Sister," and higher clergy as "Father." Genetic mutants, whose possession of immortal souls and thus of the right to live is secured by papal edict, are known as "Pope's nephews" or "Pope's children."[32] A ship leaving earth for another star system will establish an "independent daughter house" of the Order, and once that house sets up its Patriarchal See it will become, in turn, "a mother house of the Visitationist Friars of the Order of Saint Leibowitz of Tycho."[33] The cumulative effect of these metaphors is not to consecrate the biological descent that grounds them but rather to populate the novel so densely with alternative, often elective family structures that biological kinship is remanded to the margins, occupying the foreground only through analogy. On the margins, the procreative figures are not human ones but the vultures who appear at the ends of books one and two, laying their eggs in season and lovingly feeding their young, often on the meat of the novel's more hapless human characters. These carrion birds allow the narrator to satirize anthropocentrism ("[The buzzards'] philosophers demonstrated by unaided reason alone that the Supreme *Cathartes aura regnans* had created the world especially for buzzards"),[34] but they are not figures through which biological generation is equated with death. Rather, they establish predation and procreation as both baseline and limit: as constant processes that are preconditions for narration and for the creation and maintenance of the symbolic order without being themselves narratable. We might think of the buzzards as hospice figures for narrative in *A Canticle*, accompanying both human characters and narrative epochs to darkness even as they hold the novel's procreative line. That the last buzzards we see, in the novel's final pages, are likely dying of radiation poisoning even as they wait for a human protagonist to die indicates the dire pass to which the second Flame Deluge has brought life on earth. The novel's final image is neither of human death nor of the archival safe-house but of the disruption of the marine food chain: "The ash fell into the sea and into the breakers. The breakers washed dead shrimp ashore with the driftwood. Then they washed up the whiting. The shark swam out to his deepest waters and brooded in the old clean currents. He was very hungry that season."[35] All while the Memorabilia are hurtling into space to become the treasure of a daughter house that will become a mother house.

We might expect a novel with so dispassionate a view of its protagonists, so longitudinal a view of biological, symbolic, and institutional processes, to be uninterested in the ethical stakes of a single death. Brother Francis, one of the central figures of the first section, meets his end in two lines—"The arrow hit him squarely between the eyes. / '*Eat! Eat! Eat!*' the Pope's child cried"—and the philoprogenitive vultures will descend to finish what the cannibal mutant leaves behind.[36] If *A Canticle for Leibowitz* were interested solely in the perpetuation of the monastic Order and its Memorabilia, individual lives would be valuable only insofar as they contributed to that end. But Miller's novel, for all its detachment in relating death, avoids instrumentalizing its characters to the project of the monastic organism. Reproductive futurism subordinates the life of the adult to that of the child, justifying the sacrifice of the parent in the name of the child's, the genome's, and thus the future's continuity. In an extraordinary scene in *Canticle*'s middle book, abbot Dom Paulo thinks ruefully of "us monastic ignoramuses, children of dark centuries, many, entrusted by adults with an incomprehensible message, to be memorized and delivered to other adults." Yet this figuration of the child as present-tense messenger rather than future-tense messiah does not evacuate the present of its ethical importance. As he begins to suffer from the illness that will eventually kill him, Dom Paulo thinks, "Does the chalice have to be now right this very minute Lord or can I wait awhile? But crucifixion is always now. Now ever since before Abraham even is always now."[37] "Are you ready to get nailed on it?" Abbot Zerchi, Dom Paulo's book three counterpart, asks a younger monk apropos of a plan to send the Memorabilia away from Earth in the event of a second Flame Deluge. And the younger man, looking for guidance during his Gethsemane, thinks, "Destiny always seems decades away, but suddenly it's not decades away; it's right *now*. But maybe destiny is always right now, right here, right this very instant, maybe."[38] Linear time entrains the present in a consequentialist flow, subordinating the decision to its authorizing origins or to its impact on a future seen putatively in advance. Rather than rely on a past crucifixion that redeems all subsequent sin, or on a future child, messiah, or Integrator who will justify present suffering, *A Canticle* loads the present with the stakes of a perpetual crucifixion.

This ethical saturation of present suffering becomes the ground of the novel's final dilemma, one that occupies much of book three. After a limited retaliatory strike obliterates the nearby city of Texarkana, Zerchi's monastery becomes the site of a public health station that treats radiation victims and offers to euthanize and cremate those exposed to supercritical dosages. Although this state-sanctioned euthanasia requires due legal process, Abbot Zerchi opposes it on the grounds that the Catholic Church forbids suicide under any circumstances. The ensuing arguments Zerchi has with a doctor from the aid station and with an irradiated woman seeking euthanasia for herself and her child look like a simple standoff between Church and state, God and Caesar. But in circuit with the notion of perpetual crucifixion, these arguments can be seen to entail rival *temporalities* more than rival sets of permissions and interdictions. As Zerchi observes to an aid station doctor, any state that makes euthanasia provisions in advance of the nuclear war that would activate them presupposes either its own commission of a war crime or its catastrophic failure to avert the war crime of another nation. "[I]nstead of trying to make the crime impossible, they tried to provide in advance for the consequences of the crime," says Zerchi. To the doctor's assertion that mercy-killing is better than condemning victims to a slow, painful death, Zerchi responds, "Is it? Better for whom? The street cleaners? Better to have your living corpses walk to a central disposal station while they can still walk? Less public spectacle? Less horror lying around? Less disorder? A few million corpses lying around might start a rebellion against those responsible. That's what you and the government team mean by better, isn't it?"[39] The state produces its futurity not by a commitment to spare its citizens from radiation exposure at all costs but by making provisions to cut short, in the name of mercy, the agonizing and politically disruptive futures of the citizens it fails to protect. Mercy thus becomes a license for perpetuating the state through war, setting the state's future in an inverse relation with that of its injured citizens. Against this weaponization of mercy, with its chilling applicability to the euthanizing or sterilizing of disabled and sexually dissident subjects, Zerchi's exhortation to the sick woman—to pray in the face of pain rather than yield to "despair, anger, loss of faith… [or the] false god of expedient mercy"—amounts to a queer temporality: a replenishment of the present, the crucifixion that is always now, in the face of the warfare state's attempt to territorialize the future.[40] To read *A Canticle for Leibowitz* for its resistant temporalities is to find the novel queerest where it is most Catholic.

After a losing altercation with the aid station guards, Abbot Zerchi returns to the monastery to hear the confession of a bicephalous mutant. This is Mrs. Grales the tomato-seller, who has been seeking baptism for her childlike, seemingly vestigial second head, whom she calls Rachel. But the "genetic festering" caused by the war eighteen centuries ago continues to blur distinctions between generations and even between ensouled beings, to the confusion of the Church; as Zerchi puts it, "There's some question as to whether Rachel is her daughter, her sister—or merely an excrescence growing out of her shoulder."[41] Mrs. Grales surprises her confessor by announcing her desire to forgive "Him who made me as I am… Mayn't an old tumater woman forgive Him just a little for His Justice?" But before her confession is complete, a blinding flash announces the arrival of a full-scale nuclear exchange, bringing the church down around Zerchi, Mrs. Grales, and Rachel. Pinned in the rubble, the abbot finds the collapse has broken open the crypts and disgorged the bones of his forerunners, including a skull with a hole in the forehead where the remains of an arrow are lodged. "Brother," says Zerchi to what may be the skull of Brother Francis, and as he blesses it the two celibates touch across time, each having played a role in preserving the Memorabilia. Far from delegitimizing their work for having transmitted the deadly viruses of literacy and technology, the tender communion of the dying man with the dead suggests their vigil over the remains of the archive was no less worth keeping for having produced another Flame Deluge; that the future was not foreordained but open; that in keeping watch, giving Christ to human beings, and warning them that culture "could never be Eden" they had done what was given them to do.

The last stage in the Passion of Abbot Zerchi is the strangest, bringing home how dissident temporalities both prop up linear time and bloom in its ruins. Although Mrs. Grales's primary head seems dead or asleep owing to the blast, the radiation seems to have wakened the Rachel-head, who now repeats Zerchi's words in a childlike voice and looks on him with curious eyes and "a young shy smile that hoped for friendship," her head supported by the suddenly rejuvenated body of the old tomato-seller. Zerchi tries to fulfill Mrs. Grales's wish of baptizing Rachel but the latter recoils, then picks up the ciborium he had been holding before the church collapsed and offers him, without conventional words or gestures but "as if by *direct instruction*," the Host. "Now he knew what she was, and he sobbed faintly when he could not again force his eyes to focus on those cool, green, and untroubled eyes of one born free."[42] Earlier his fellow monk Joshua had dreamed in a "blasphemous nightmare" of Rachel saying, "Accurate am I the exception… I commensurate the deception. Am… I am the Immaculate Conception."[43] The second head of a bicephalous woman, born of one nuclear war and awakened by another, daughter or sister or excrescence or co-vivant of Mrs. Grales, is the second coming of the Virgin Mary, conceived without stain of original sin and therefore needing no baptism. Even as linear time is spiraling back on itself with the arrival of the second Flame Deluge, the miracle of Rachel—"Accurate am I the exception"—takes place at the crossroads of aberrant temporalities: cyclicality, parthenogenesis, innocence in apocalypse, wakening amid death. The obverse of the carrion birds, she escorts Zerchi from his life under the signs of miracle, sacrament, transubstantiation. Her cool touch on his forehead is the last thing he feels; hers is the last word he will hear before he dies: "Live."

The final chapter of *A Canticle for Leibowitz* opens with a rare glimpse of children. These are boarding a spaceship bound for an existing colony in a neighboring star system: "They sang as they lifted the children into the ship. They sang old space chanteys and helped the children up the ladder one at a time and into the hands of the sisters. They sang heartily to dispel the fright of the little ones. When the horizon erupted, the singing stopped. They passed the last child into the ship."[44] As the ship departs against the backdrop of the second Flame Deluge, the echoes of Noah's ark are clear. But instead of ferrying the animals two-by-two—the procreative minima of species regeneration—to the Alpha Centauri colony, this ark carries monks, nuns, and orphans; a few scientists and staff; three bishops, who alone will be able to ordain more priests; the Chair of Peter so that the next Pope can be crowned off world; and, on microfilm, the Memorabilia. All this as part of *Quo pereginratur grex, pastor secum* ("Whither wanders the flock, the shepherd is with them"), an emergency plan for "perpetuating the Church" on another world should Earth become uninhabitable.[45] A one-species ark without animals, the *Quo peregrinatur* ship is less concerned with perpetuating *Homo sapiens* than with ensuring the survival of a single transtemporal organism, the Church. [46] In concert with the theologically informed miracle of Rachel, this final emphasis might seem to reorient *A Canticle* away from the preservation of the Memorabilia and toward that of Catholicism. By such a reading, the Memorabilia would be preserved only as a means to the end of securing the Church's worldly power and spiritual authority. But while the novel's middle chapters feature the odd abbot leveraging possession of the Memorabilia for political advantage, they more often show us how the monastery is exposed by its archive to risk of attack, to ideas that run counter to the Order's doctrine, and to hi-tech secular powers that owe their rise to the Memorabilia but, having exploited it, no longer need to consult it or to kowtow to its guardians. In preserving the Memorabilia, the Church carries the seeds of its loss of monopoly on metaphysics and its prospective obsolescence as a worldly power. Yet like Rachel, the other child who bears a message to an adult, the Church is awakened from this dormancy by the self-destruction of worldly technological states whose impatience for earthly paradise—the Eden Zerchi and his predecessors warned against expecting in this life—is their own death drive. In Miller's novel, church and state are braided in a chiasmic chain, one waning while the other waxes, each with a death drive lodged, paradoxically, at the very heart of its plans for self-perpetuation.

A Canticle for Leibowitz attributes a death drive not only to church and state but also to the archive itself, an ineliminable tendency in writing to stockpile and disseminate the prospective conditions of its effacement: it is thanks in part to the Memorabilia—not just the particular holdings of the abbey but the techno-scientific archive generally—that the Memorabilia are imperiled and must be conveyed off the planet. In this respect, as in so much else, Miller's *mise-en-scène* anticipates Derrida, who would theorize what he variously names "archive fever," the "death drive of the archive," the "archiviolithic force," and "the violence of the archive itself, *as archive, as archival violence.*" "It is at work," he writes in *Archive Fever* (1995), "but since it always operates in silence, it never leaves any archives of its own. It destroys in advance its own archive, as if that were in truth the very motivation of its most proper movement. It works *to destroy the archive: on the condition of effacing* but also *with a view to effacing* its own 'proper' traces."[47] A decade after "No Apocalypse, Not Now," Derrida's master-figure for the immanent violence of the archive remains the nuclear. He refers repeatedly to the archive as "a series of cleavages [that] will incessantly divide every atom of our lexicon"; as "haunted from its origin" by "the possibility of its fission"; as haunted, too, by secrets that are "the very ash of the archive."[48] For Derrida, the fission of the word is inherent in its inscription, the effaceability of writing its indispensable precondition. This fissibility of the word inheres at every possible scale: in the cleft etym, divided against itself by its lexically unstable components and its capacity for resignification; in the catastrophes, both slow and split-second, that destroy writing's material and digital traces; in the linguistic drift that can efface the intelligibility of writing even when it survives; in writing's way of making my death and absence present to me. But if indeed the archiviolithic destroys in advance its own archive, we cannot point to it, cannot even point to its remains. We can only speculate about it as we speculate about death: hence the strange compatibility of the archiviolithic, the archival death drive, with speculative fictions of the nuclear, the event that has so far occurred, at its fullest blown, only in discourse, as a purely theoretical event.

Neither Miller's speculative fiction nor Derrida's work on the archive addresses queer sex head-on, although celibacy, a major element in *Canticle*, is currently being rethought as a queer sexuality—rather than, say, as the absence of sexuality, or as closeted homosexuality—in its relation to reproductive time.[49] Bringing queer temporalities frameworks into conversation with Nuclear Criticism will certainly give us new ways to think about nuclear conditions both pre- and post-1989 in conjunction with queer sex and sexuality specifically. What Patrick R. Mullen calls the "vehicular mobility" of the term *queer* "as a capacious index for a series of non-normative desires, sexualities, people, politics, and cultural expressions" will also allow us to consider forms of dissidence that maintain, say, a temporal connection to queer sexuality even when the latter does not figure, or figures obliquely, in a work's diegesis or propositional content.[50] This second kind of work, which I have tried to undertake here, can crucially spoil the non-stick surface of "history," which presents itself as the only authorized finish for frying the egg of time. Resisting even the notion of a queer theoretical "turn toward time" as conceding too much to "the motionless 'movement' of historical procession obedient to origins, intentions, and ends whose authority rules over all," Lee Edelman would pose against it the life-in-death—or we might say, echoing Miller, the crucifixion that is always now—of the drive's cussedness and recursivity, which the queer subject is made to figure.

> Whether polyphonous or univocal, history, thus ontologized, displaces the epistemological impasse, the aporia of relationality, the nonidentity of things, by offering the promise of sequence as the royal road to *con*sequence. Meaning thus hangs in the balance—a meaning that time, as the medium of its advent, defers while affirming its constant approach, but a meaning utterly undone by the queer who figures its refusal. This is the truth-event, as Badiou might say, that makes all subjects queer: that we aren't, in fact, subjects of history constrained by the death-in-life of futurism and its illusion of productivity. We're subjects, instead, of the real, of the drive, of the encounter with futurism's emptiness, with negativity's life-in-death. The universality proclaimed by queerness lies in identifying the subject with just this repetitive performance of a death drive, with what's, quite literally, unbecoming, and so in exploding the subject of knowledge immured in stone by the "turn toward time."[51]

The subject, Edelman insists, is not who overcomes the death drive but who is identified with it, and whose universal queerness inheres in this identification with repetition and the traumatic real. Such a subject stands in a queer and unassimilable relation to historical narratives grounded in procession, continuity, and reproductive futurism. A Nuclear Criticism that heeded an insistence like Edelman's would write against time's collapse into history; against the script that says our children are safe now that nuclear weapons are no longer such a threat; and against the heternormative logic that exceptionalizes queer subjects as non-reproductive bearers of the death drive. And it would wrest from history's chronologizing authority the dissident temporalities that get subjected to it: cyclical, uncanny, and preposterous ways of being in time; discontinuous and transtemporal ones; births without conception and second comings that are at once awaited and unprecedented; and those evidently barred futurities in whose face we make some of our severest refusals to unbar the future on deplorable terms. Such a criticism would think the nuclear condition not only in numbers of warheads, ages of missiles, degrees of alert, or locations of fissile material but also as metonymizing and materializing the death drives of church, state, archive, and subject. It would recognize that we might dismantle every nuclear weapon on the planet—as we should—without coming close to dismantling the nuclear condition. To a history that placed the nuclear condition firmly in the past, *back then*, such a criticism would respond: the nuclear condition, like crucifixion, is always *now*.

Notes

[1] I would like to thank Patrick Moran and the editors of this volume for commenting on drafts of this essay.
[2] Jonathan Schell, *The Fate of the Earth: and The Abolition* (Stanford: Stanford University Press, 2000), 154, 119.
[3] Ibid., 157-58.
[4] James's scenario, although unkeyed to a nuclear event, bears a striking resemblance to one Schell envisions as a sequel to a survivable nuclear exchange: "It is possible…at least to imagine that, through sterilization of the species, the future generations could be cancelled while the living were left unharmed…They would experience in their own lives the breakdown of the ties that bind individual human beings together into a community and a species, and they would feel the current of our common life grow cold within them. And as their number was steadily reduced by death they would witness the final victory of death over life." Ibid., 168-69.
[5] By the phrase "nuclear condition," I do not mean the simple existence of nuclear weapons. Even if the last nuclear weapon were dismantled the nuclear condition would persist, inhering in our knowledge that nuclear weapons cannot (short of a knowledge-loss even more catastrophic than the one imagined by *A Canticle for Leibowitz*) be uninvented, that our history has been irreversibly shaped by nuclear weapons, and that future arsenals might again contain them. I also mean the phrase to reactivate, without fully embracing, Derrida's argument that literature has been "nuclear" since the Enlightenment insofar as it depends for its meaning on an archive susceptible to physical destruction, and that inscription is therefore premised on the possibility of its own effacement. Finally, the notion of a nuclear *condition* is counterposed to the nuclear *event* whose prospective arrival nonetheless defines the nuclear condition.
[6] This rubric was consolidated in a special Queer Temporalities issue of *GLQ: A Journal of Lesbian and Gay Studies* 13, no. 2-3 (2007), edited by Elizabeth Freeman. The issue's opener, "Theorizing Queer Temporalities: A Roundtable Discussion" (pp. 177-95) brought together nine of the critics most often associated with this area of study (Carolyn Dinshaw, Lee Edelman, Roderick A. Ferguson, Carla Freccero, Elizabeth Freeman, Judith Halberstam, Annamarie Jagose, Christopher Nealon, and Nguyen Tan Hoang).
[7] The expression "reproductive futurism" is developed in Lee Edelman, *No Future: Queer Theory and the Death Drive* (Durham, NC: Duke University Press, 2004).
[8] Carolyn Dinshaw, *Getting Medieval: Sexualities and Communities, Pre- and Postmodern* (Durham, NC: Duke University Press, 1999), 151; Dinshaw et al., "Theorizing Queer Temporalities," 190.
[9] See also Ann Cvetkovich, *An Archive of Feelings: Trauma, Sexuality, and Lesbian Public Cultures* (Durham, NC: Duke University Press, 2003); and the

work of Melanie Micir, particularly "'Living in Two Tenses': The Intimate Archives of Sylvia Townsend Warner," forthcoming in *Journal of Modern Literature*; and Micir's doctoral dissertation, "Public Lives, Intimate Archives: Queer Biographical Practices in British Women's Writing, 1928-1978."

[10] See, respectively, Leo Bersani, "Is the Rectum a Grave?" *October* 43 (Winter 1987): 197-222 and *The Culture of Redemption* (Cambridge, MA: Harvard University Press, 1990); Edelman, *No Future*; and Heather K. Love, *Feeling Backward: Loss and the Politics of Queer History* (Cambridge, MA: Harvard University Press, 2007).

[11] See Judith Halberstam, "The Anti-Social Turn in Queer Studies," *Graduate Journal of Social Science* 5, no. 2 (2008): 140-56.

[12] Compare: "The only hopeful and sustainable future for our children, the cities and people of this world is a nuclear weapons free future" (Tamara Lorincz, Halifax Peace Coalition, 2007); and "While nuclear weapons exist, we cannot leave ourselves and our children open to the threat of nuclear blackmail" (UK Liberal Democratic Party's "shadow defense review consultation paper," 2012). See *Welland Tribune* (August 3, 2007) http://www.wellandtribune.ca/ArticleDispl ay.aspx?archive=true&e=638569; and Richard Norton-Taylor, "UK's Nuclear Weapons in the Frame," *Guardian* (March 6, 2012) http://www.guardian.co/uk/ news/defence-and-security-blog/2012/mar/06/uk-nuclear-trident. The point here is not that the nuclear weapons debate has a monopoly on reproductive futurism. As a security question with long-term geopolitical and environmental ramifications, however, the nuclear weapons debate deploys reproductive futurism in particular ways and with special intensities that should be distinguished from the general invocation of "our children and grandchildren" as the chief beneficiaries of, and thus the master rationale for, policy-making.

[13] Jacques Derrida, "No Apocalypse, Not Now (full speed ahead, seven missiles, seven missives)," trans. Catherine Porter and Philip Lewis, *diacritics* 14, no. 2 (Summer 1984): 20-31; Susan Sontag, *AIDS and Its Metaphors* (New York: Farrar, Straus, and Giroux, 1989), 88.

[14] Peter Coviello, "Apocalypse From Now On," in *Queer Frontiers: Millennial Geographies, Genders, and Generations*, ed. Joseph A. Boone et al. (Madison, WI: University of Wisconsin Press, 2000), 42.

[15] Ibid., 45, 49; original emphasis.

[16] *Pace* Coviello, the chronology of the 1980s suggests a narrative of overlap and feedback rather than of strict succession between Nuclear Criticism and queer theory in the HIV age. AIDS was named during the same year in which *The Fate of the Earth* was published, and HIV identified as its cause in 1984, the year of the Nuclear Criticism conference at Cornell University.

[17] See Paul K. Saint-Amour, "Bombing and the Symptom: Traumatic Earliness and the Nuclear Uncanny," *diacritics* 30, no. 4 (Winter 2000): 59-82.

[18] Derrida, "No Apocalypse, Not Now," 20.

[19] Ibid., 24.

[20] Ibid., 26.

[21] Ibid., 28.

[22] Ibid., 30-1.
[23] Ibid., 31.
[24] Ibid., 31.
[25] See Michael O'Rourke, "Queer Theory's Loss and the Work of Mourning Jacques Derrida," *Rhizomes* 10 (Spring 2005), http://www.rhizomes.net/issue10/orourke.htm; and the forthcoming *Derrida and Queer Theory*, ed. Michael O'Rourke (Basingstoke and New York: Palgrave Macmillan, 2013).
[26] Derrida writes: "[O]n the topic of this name, 'nuclear criticism,' I foresee that soon, after this colloquium, programs and departments in universities may be created under this title, as programs or departments of 'women's studies' or 'black studies' and more recently of 'peace studies' have been created... ." In Derrida, "No Apocalypse, Not Now," 30.
[27] Ibid., 27.
[28] Walter M. Miller, Jr., *A Canticle for Leibowitz* (New York: HarperCollins, 2006), 62, 26; original emphasis. Miller was a tail gunner and radio operator in U.S. Army Air Force bombers during the Second World War, and in 1944 his crew flew in the missions that destroyed the abbey at Monte Cassino, Italy, the oldest monastery in the West and the site where St. Benedict wrote the *Regula Benedicti*. Allied commanders were convinced—wrongly, it turned out—that the abbey was being used by German forces as an artillery observation post. Fortunately, two German officers had arranged the previous year for the priceless contents of the abbey's archives, library, and gallery to be evacuated to the Vatican. Miller evidently did not realize until he was drafting *A Canticle*'s penultimate scene that the novel was his response—perhaps, in part, a reparative one—to his wartime experience.
[29] Ibid., 65.
[30] The novel's three books are set in 2560 C.E. ("Fiat Homo"), 3174 C.E. ("Fiat Lux"), and 3781 C.E. ("Fiat Voluntas Tua").
[31] Miller, *A Canticle for Leibowitz,* 273.
[32] Ibid., 96.
[33] Ibid., 282, 289.
[34] Ibid., 239.
[35] Ibid., 334.
[36] Ibid., 114.
[37] Ibid., 153-54.
[38] Ibid., 282, 286.
[39] Ibid., 292-93.
[40] Ibid., 314-15.
[41] Ibid., 275, 272.
[42] Ibid., 330, 332.
[43] Ibid., 276.
[44] Ibid., 333.
[45] Ibid., 266.
[46] One important transtemporal figure I don't discuss in the body of the essay is the hermit Benjamin, or Eleazar, the one character who seems to persist through all

three books of *A Canticle*. His age (he claims to be 3,209 in the year 3174 C.E.) would make him the same Lazarus resurrected by Christ, as if having been raised from the dead he could not then die. Benjamin appears to be the last Jew in the Catholic world of the novel; his longevity associates him with the medieval figure of the Wandering Jew, who taunted Jesus on the Via Dolorosa and was cursed to wander the earth until the second coming. Despite Benjamin's status as the narrative's exception (to mortality, to Christianity), his persistence across millennia allies him with the novel's other figures and forms of recalcitrant temporality. For an engaging discussion of Benjamin and of the broader importance of Jewishness for Miller's account of the abbey and its mission, see Amy Hungerford, *The Holocaust of Texts: Genocide, Literature, and Personification* (Chicago: University of Chicago Press, 2003), 67-71.

[47] Jacques Derrida, *Archive Fever: A Freudian Impression*, trans. Eric Prenowitz (Chicago: University of Chicago Press, 1996), 10. The essay was first read at a colloquium in London in 1994.

[48] Ibid., 1, 100.

[49] See, for example, Benjamin Kahan, "'The Viper's Traffic Knot': Celibacy and Queerness in the 'Late' Marianne Moore," *GLQ* 14, no. 4 (September 2008): 509-35; and *Celibacies: American Modernism and Sexual Life* (Durham, NC: Duke University Press, forthcoming).

[50] Patrick R. Mullen, *The Poor Bugger's Tool: Irish Modernism, Queer Labor, and Postcolonial History* (New York: Oxford University Press, 2012), 6.

[51] Lee Edelman comment, in Dinshaw et al., "Theorizing Queer Temporalities," 181.

CHAPTER FOUR

APOCALYPSE NETWORKS:
REPRESENTING THE NUCLEAR ARCHIVE

BRADLEY J. FEST

> I considered fire, but I feared that the burning of an infinite book might be similarly infinite, and suffocate the planet in smoke.
> —Jorge Luis Borges, "The Book of Sand"[1]

Introduction: The Big Red Button and the Kill Switch

During the summer of 2010 Senator Joseph Lieberman proposed a bill for an Internet "kill switch," a bill that would grant the President of the United States the "far-reaching emergency powers to seize control of or even shut down portions of the Internet."[2] Though Lieberman quickly qualified the reach of his Protecting Cyberspace as a National Asset Act (PCNAA), stressing the relatively limited control it would grant the President,[3] the idea of an Internet kill switch should give one significant pause in terms of the nuclear imagination of the twentieth century. The metaphor of the kill switch, a singular button or device that gives the President of the United States instantaneous control over a significant portion of a vast and powerful network, is a trope whose origins clearly reside in the Cold War's semi-mythical "Big Red Button." The prospect of Mutual Assured Destruction (MAD), the dominant U.S. national fantasy of the mid-twentieth century, was often understood in the popular imagination as potentially resulting from a single action. And indeed, the instantaneity and abstraction of global nuclear war necessitated such a singular metaphor in order for the massive systemic complexity of the North American Aerospace Defense Command (NORAD) and its global capabilities to be comprehended. But the technological, political, and economic realities surrounding nuclear capability are simply too complex to be captured by The Button in precisely the same way that Lieberman's bill is a fantasmatic and reactionary attempt to explain and control an even more complex system than NORAD. The Internet, like the international capability of waging nuclear war, is simply too systemically and asymmetrically massive to be controlled in such a singular fashion.

Among other things, the metaphors of the Big Red Button and the Internet kill switch emerge from an encounter with what Fredric Jameson famously calls the postmodern sublime. These two metaphors should be seen as "degraded attempts—through the figuration of advanced technology—to think the impossible totality of the contemporary world system."[4] The button and the switch stand in for totalities, for the vast networks of communication they control and the catastrophic potential destruction they represent. These systems are too large and complex to be "cognitively mapped." Consequently, they are often represented in popular expression as singular metaphors that only vaguely correspond to the reality they represent, reifying the totalities that they stand in for, while their symbolic value and speculative projection often have very serious real world consequences.[5] Even though Lieberman's bill died fairly quickly on the floor of Congress, it is representative of the power and persistence of the nuclear referent to shape juridical practice and political discourse (even in the perceived *absence* of that referent).

Thus the PCNAA, like the other twenty-first century speculative projections of archival destruction that I will be exploring here, should be read as a significant and telling manifestation of the nuclear imagination of the present age. The speculative eschatological narrative of the United States' power over instantaneous material destruction on a nearly unimaginable, species-wide scale has transformed in the wake of the Cold War, the attacks of September 11th 2001, the subsequent War on Terror, the global emergence of Internet 2.0, the global financial crisis, and observable climate change into a fantasmatic projection of informational control and ecological catastrophe. Like MAD, the material consequences of deploying an Internet kill switch, with even a slight amount of imaginative extrapolation, would clearly be disastrous considering the global reliance on the Internet for a great many of the species' economic, political, social, biological, and ecological activities.

Further, there are interesting parallels to be drawn between the speculative narratives implicit in the PCNAA and the nuclear imagination. Recall that the Internet itself was largely an outgrowth of the perceived threat of MAD and the Advanced Research Projects Agency (ARPAnet). As media theorist Alexander Galloway notes in *Protocol*: "While many have debated the origins of the Internet, it's clear that in many ways it was built to withstand nuclear attack. The Net was designed as a solution to the vulnerability of the military's centralized system of command and control during the late 1950s and beyond. For, the argument goes, if there are no central command centers, then there can be no central targets and overall damage is reduced."[6] The irony of Lieberman's bill is considerable. By making one system more structurally complex in order to stave off an imagined disaster, a system has emerged that is too complex to control and politicians are now attempting to legislate the capability to destroy or disable that original system in case of disaster, terrorism, or political unrest. Though a gross oversimplification, this, along with other reasons, is why Galloway further suggests that, "If one can consider nuclear attacks the most highly energetic, dominating, and centralized force that one knows—an archetype of the modern era—then the Net is at once *the solution to and the inversion of* this massive material threat, for it is precisely noncentralized, nondominating, and nonhostile."[7] Cybernetic and nuclear technologies, who share their origins in significant American and British military efforts during the Second World War,[8] should thus be read as structurally and historically intertwined.[9] A Nuclear Criticism following the Cold War cannot ignore the imaginative and historical forces produced by the continued dialogue between information and military technologies, between the archive and the Bomb, between the decentralization of the first nuclear age and the networked distribution of the second age in which the nuclear referent has dispersed in a variety of ways, some of which I will be attending to in this chapter.[10]

Among the many crucial insights of Jacques Derrida's famous essay "No Apocalypse, Not Now" (1984) is his emphasis that fantasies of nuclear destruction should primarily be understood as projections of archival destruction:

> Now, what the uniqueness of nuclear war, its being-for-the-first-time-and-perhaps-for-the-last-time, its absolute inventiveness gives us to think, even if it remains a decoy, a belief, a [f]antasmatic projection, is obviously the possibility of an irreversible destruction, leaving no traces, of the juridico-literary archive and therefore of the basis of literature and criticism. [. . .] The hypothesis we are considering here is that of a total and remainderless destruction of the archive.[11]

Lieberman's PCNAA, which at the time of this writing remains a fantasy, follows from this critical hypothesis and defines the eschatological limit of the archive along a different, yet parallel line; the contemporary archive *par excellence*, the Internet, here becomes capable of destroying itself, a capability that depends upon its underlying nuclear logic. And its disappearance would accompany some catastrophe or else would be catastrophic itself. During the second nuclear age the hypothesis of total archival destruction, though initially given to thought by the possibility of nuclear war, sheds its initial nuclear trappings and the destruction of the archive is given to thought *by nothing outside of archival processes themselves.*[12]

To map this transformation from the archive threatened by global thermonuclear war to the archive's threat to itself deserves a significant amount of attention, as the frequent juxtaposition of nuclear war and archival destruction in literary texts during the twentieth century more than demonstrates. Literature has long been fascinated by and anxious about its material disappearance and erasure. And yet, as Fernando Báez asks, "There are hundreds of studies on the origin of books and libraries, but there is not a single history of their destruction. Isn't that a suspicious absence?"[13] During the middle of the twentieth century these histories did in fact proliferate, but they were often set in the future and they were mostly fictional. The Bomb's imaginary, its relationship to history, and its non-event allowed (and for many writers required) narratives to project not only material destruction on an unprecedented scale, but, as Derrida suggests, archival destruction that "lack[s] any common proportion with, for example, the burning of a library, even that of Alexandria, which occasioned so many written accounts and nourished so many literatures."[14]

In this chapter I will begin to sketch a history of what I call the *tale of archival crisis* and examine a few examples of its current expression. Cultural artifacts that write the disaster have proliferated in the last sixty years, and so for this reason I can only trace a rough outline of the nuked archive, and my examples are less exhaustive than particular. But even this archival accumulation and our inability to access all of it proves my thesis. If during the first nuclear age, following Derrida, the total and remainderless destruction of the archive represented the disastrous asymptotic limit of global nuclear war, the second nuclear age's eschatological limit, both in reality and literature, can be defined by the threat posed by the Bomb *and* the archive. The archival imagination of the second nuclear age not only understands the archive as system with the capability of destroying itself, a haunting possibility that would have far reaching and disastrous effects upon the world, but we can now perceive that archival *accumulation*, the underlying logic of the contemporary archival impulse to collect, store, and document everything, has the potential to be eschatologically threatening.

Tales of Archival Crisis in the First Nuclear Age

Though the representation of archival destruction has an obviously rich literary tradition, and critics such as Paul K. Saint-Amour[15] have masterfully demonstrated the presence of the nuclear referent in texts written before 1945,[16] three notable early Cold War novels—George R. Stewart's *Earth Abides* (1949), Walter M. Miller Jr.'s *A Canticle for Leibowitz* (1956), and Nevil Shute's *On the Beach* (1957)—readily suggest themselves as points of departure. These novels are particularly representative of the early nuclear imagination, having long been understood as important manifestations of Cold War nuclear anxieties. Further, they are exemplary sketches of three distinct and important archival limits. These limits provide a structure to what we can (retroactively) see as aspects of the first nuclear age's relationship to the archive.

Though the apocalypse it imagines is epidemiological rather than nuclear, it is impossible to read Stewart's *Earth Abides* without a sense of the historical fallout from the bombing of Hiroshima, which occurred only four years previous to the novel's publication. *Earth Abides* tells the story of a small group of survivors on the West Coast of the United States. One of the central threads of the novel follows Ish, the protagonist, and his urge to save an old library from destruction. When his son Joey shows a similar interest in knowledge, Ish begins taking him to the library. Joey's tragic death ultimately ends Ish's desire to protect the library, it falls into ruin, and the novel ends in a future where the species is unable to read its own past. The archive may exist in some form at this particular limit of the eschatological archive, but it is illegible in Stewart's speculation.

Shute's *On the Beach* similarly ends with the archive's preservation (though this occurs tangentially). In the wake of a global nuclear war, extreme enough to poison the entire planet with radiation but not extreme enough to destroy *everything* (e.g., there is a moving scene in which the characters encounter a preserved yet irradiated Seattle), the novel narrates a humanity awaiting its inevitable end. Here the archive is preserved but there is no one around to read it.

Miller, in *A Canticle for Leibowitz*, traces a slightly different path. Set in a monastery in a post-apocalyptic Southwestern United States, the first part of the novel concerns a scribe's effort to preserve and illuminate a pre-nuclear manuscript. This manuscript, the "blessed documents"[17] whose illumination is of great importance to Brother Francis, ironically turns out to be nothing more than mundane blueprints. This realization and the action of the rest of the novel emphasize the archive as a site of misreading, stressing that a hermeneutics of the archive, particularly a post-nuclear archive, will necessarily involve misinterpretation.[18]

These three limits roughly encapsulate the boundaries of the archival imagination produced by the projection of nuclear war as it was expressed during the first nuclear age. In the speculated wake of MAD there will either be no-one around to read the archive, literacy will be lost, or else the past will be quite difficult to decipher. And in a sense, these limits are not unique in the historical literary imagination.[19] For, as in earlier texts like Nathaniel Hawthorne's "Earth's Holocaust" (1844) or William Carlos Williams's *Spring and All* (1923), in each of these scenarios, something *remains*. Historian Carolyn Steedman usefully calls this remainder "dust": "[Dust] is not about rubbish, nor about the discarded; it is not about a surplus, left over from something else: *it is not about Waste*. Indeed, Dust is the opposite thing to Waste, or at least, the opposite principle of Waste. It is about circularity, the impossibility of things disappearing, or going away, or being gone. Nothing *can be* destroyed."[20] The archive in each novel is not totally destroyed but remains in some form or another after the narrative has ended as *dust*.

Obviously then, there is a fourth limit that is clearly absent here: Derrida's total and remainderless destruction of the archive. If the nuclear referent for the first time provides the specter of total bibliocaust, we must realize simultaneously that the success of projecting such an abyssal proposition through mimesis is doomed.[21] For there to be anything constructed or expressed in the face of archival destruction of this magnitude, even some irradiated dust must remain, something to ground, however shakily, the act of representation. As literary critic James Berger in his compelling study of twentieth century post-apocalyptic narrative and discourse suggests, "in nearly every apocalyptic presentation, something remains *after the end*,"[22] and indeed, there is "a pervasive post-apocalyptic sensibility in recent American culture. It seems significant that in the late twentieth century we have had the opportunity [. . .] to see after the end of our civilization—to see in a strange prospective retrospect what the end would actually look like: it would look like a Nazi death camp, or an atomic explosion, or an ecological or urban wasteland."[23] And a great deal of modern and postmodern cultural production, perhaps nowhere more evident than in twentieth century nuclear narratives, has expanded our perspective on this post-apocalyptic present.

The nuclear archives of Stewart, Shute, and Miller are born from the trauma of a post-apocalyptic archive fever: a desire to preserve the present after the end coupled with a despair that one could never hope to read, interpret, or decipher the post-nuclear archive. To put it broadly, projecting archives of disaster fundamentally shaped our conception of archives of all kinds during the first nuclear age, and we quite quickly imagined archival processes as defense mechanisms against future devastation. The nuclear bomb also provided the archive with a definite *telos*, a reason for existing that it previously lacked in such stark terms. The *intentional* archive as imagined in Arthur C. Clarke's *Rendezvous with Rama* (1973), and Larry Niven and Jerry Pournelle's *The Mote in God's Eye* (1974), for example, becomes an archive whose explicit goal is to preserve aspects of civilization in the wake of its collapse. Archives have always been about preservation, about cataloging and organizing past documents for future users, but the archive's *raison d'être* was historical, progressive, self-justifying, and tautological (i.e. preserving the past for future historians was unquestionably important; the archive should be preserved because the archive should be preserved). In the nuclear imagination the archive becomes a site that anticipates its own remainderless destruction and often plans for it.[24]

Nuclear archives, whether real or fictional, produce a kind of dust *avant la lettre*. With this in mind I think it is appropriate to modify slightly Walter Benjamin's famous dictum to describe such an archive: "There is no document of civilization which is not at the same time a document of [future] barbarism."[25] The power over the nuclear archive is not only over history and the present, but over "post-history," an imagined future whose only tie to the past is through the post-nuclear archive. Quite simply, how we imagine this type of future directly shapes the archival constructions of the present. Consequently, one of the goals of a revitalized Nuclear Criticism should be to realize that one of the central debates during its brief vogue in the late-1980s to the early-1990s, the disagreements about the "real" versus the "textual" nature of nuclear war,[26] should be reframed. The first nuclear age, which came to a close with the end of the Cold War, can now be characterized by this complex interplay between history, the archive, and speculative destruction, a temporality produced by the specter of global thermonuclear war that was, as Derrida puts it, "a phenomenon whose essential feature is that it is *fabulously textual*, through and through,"[27] *and* quite real in the sense that MAD failed to occur alongside, and in a certain sense because of, the nuclear archival imagination of this period.

By imagining the various limits of archival destruction, the limit beyond which such destruction could not pass, the total and remainderless destruction of the archive has remained unrealized. Stewart, Shute, and Miller, by projecting their various archival limits into the future and letting post-nuclear dust drift from the future into the present, can now be read as inscribing the *non-event* of global nuclear war into the past. In other words, the hindsight provided by the second nuclear age is one in which MAD has moved into the archive of the past; nuclear apocalypse has become historical even though it failed to occur. Consequently, one of the many tasks given to Nuclear Critics of the second nuclear age should be to endeavor to understand the post-nuclear dust that remains, the archival inscriptions of nuclear war that continue to inflect and inform the textuality and reality of contemporaneity, disaster, and the world risk society.

Tales of Archival Crisis in the Second Nuclear Age

In the remainder of this chapter I will focus on three recent works of fiction that I believe nicely capture and complicate the archival imagination of the present: David Mamet's *Wilson: A Consideration of the Sources* (2000), Neal Stephenson's *Anathem* (2008), and Charles Stross's novella, "Palimpsest" (2009). These three texts all demonstrate that the technological master-signifier of the twentieth century, nuclear annihilation, in certain ways has been transmogrified into an informational sublime—an encounter with a massive amount of information that cannot ever be accessed in its entirety, let alone understood.[28] Each novel explicitly posits the archive in relationship to disaster, but in a much more complicated sense than, say, artificial intelligence getting out of control. Rather, these fictions all realize that in the second nuclear age archives trace complex relationships to history, temporality, and politics that cannot be easily mapped along an eschatological, linear trajectory. Nonetheless, the archive in each text remains a site of crisis in terms of archival destruction, of anticipating or reacting to nuclear war, and as an agent of disaster. These texts clearly display a relationship with the tradition of archival crisis during the first nuclear age as explored in Stewart, Shute, and Miller, while considerably reconfiguring the temporality of archival crisis. The singular and dramatic narrative event of global nuclear war, even if present at some moment in the texts' narrative history, in each of these texts is *dispersed*, both spatially and temporally. The threats Mamet, Stephenson, and Stross imagine for their archives are no longer singular, causing "irreversible destruction, leaving no traces,"[29] but have fluidly multiplied in archives that are themselves multiple, fluid, rhizomatic, and nonlinear. The singular destruction that marked the tale of archival crisis during the first nuclear age disperses throughout these texts' archives to the point that multiplicitous crisis becomes a fundamental principle of the archives' construction of themselves.

Though *Wilson*, a highly idiosyncratic novel for David Mamet,[30] is a fragmentary, experimental pastiche of text supposedly recovered hundreds of years after the Internet's destruction, it functions as an exemplary transitional text of the tale of archival crisis as it moves into the second nuclear age. *Wilson* is presented as a critical edition of an important historical text, with all the attendant scholarly notes, emendations, and commentary one would expect from such a volume. Mamet, however, is clearly parodying this mode of textual production and preservation, calling attention to the absurdity of post-archival work, the vacuity of present forms of scholarship, and the relative fragility of the institutions of twentieth century literary and cultural production.[31] Further, the speculative future it imagines is eerily similar to the one projected by Lieberman's PCNAA: the destruction of the Internet in *Wilson* leads to the collapse of civilization in a fashion hardly distinguishable from narratives of nuclear annihilation.

The novel constantly calls attention to its own act of narrating and its mode of critical analysis while simultaneously obscuring most of the relevant information necessary to understand its narrative scope. For instance, there is nothing explicit in the text of *Wilson* that informs us what purpose the fragments we have chaotically gathered in front of us serve, nor where they come from. One has to read the back cover of the book for this information: "When the Internet—and the collective memory of the twenty-first century—crashes, the past is reassembled from the downloaded memories of Ginger, wife of ex-President Wilson."[32] The novel attempts and dramatically fails to represent the "'Time of the Destruction of All Knowledge.'"[33] The late-twentieth and early twenty-first centuries in *Wilson* become a fairly brief and unrecorded period of human history: "For, dating [. . .] the Crash of the Internet in 2021, we have had a period of eighty [. . .] years of the reign of that commodity understood as 'information,' we have a scant nineteen years, the 'time of the Troubles,' before the Revelation, and the Riots."[34] The irony should be clear: for an age that worshipped the commodity of information, *Wilson*'s narrator/scholar has surprisingly little information from this period.

Wilson shares with the earlier nuclear texts I discussed above a sense of archival dust's persistence even in a world where the Internet has accidentally deleted itself for an absurd reason (a kind of *Dr. Strangelove* [1964] for the information age). Its wide use of fragmentation and historical distortion repeatedly emphasize the futility of archival reconstruction and the mimetic and hermeneutic challenges of narrating a post-infocaust history. At once hypertextual and retro-modernist, the result of the word processor and a post-Cold War literature of exhaustion, the novel materially enacts its own metafictional reconstruction of history in the wake of archival disaster.

That it does this without a nuclear explosion should be emphasized. A major difference between Mamet's novel and, say, *A Canticle for Leibowitz*, perhaps even more than the formal differences, is the apparent absence of the nuclear referent. Yet even so, the nuclear persists as an *instantaneous* technological disaster with wide-reaching global consequences here, much as instantaneous disaster remains a common feature of contemporary ecological disaster narratives (e.g., the films of Roland Emmerich, which depend upon instantaneous disaster for their narrative tension). The anxiety Mamet highlights, then, is the reverse of what it appears to be. Rather than worry about what might be lost or destroyed (indeed the novel contains a preface with the clear, if disingenuous acknowledgment that "*all* knowledge, of course, *was not lost*"[35]), Wilson paints in broad strokes the anxiety accompanying archival accumulation. Everywhere pointing to the absurdities of postmodern media-culture, contemporary archive fever, and the jargon filled minutiae of historical and literary scholarship, the novel ultimately asks not only how future historians will begin to engage with a period during which so much, perhaps too much, was saved (i.e., how does one begin to choose what to study?), but how this very approach to something like the asymptotic limit of "total knowledge" leads toward disaster. The urge to preserve the documents of the past in the face of an uncertain though eschatological nuclear future transforms into the very thing threatening that future. *Wilson* understands the present's hyper-connectivity and its encyclopedic drive as absurd, dangerous, and inevitably disastrous as the totality of this information amounts to nothing short of an absence of information.

Neal Stephenson's *Anathem* shares *Wilson*'s sense of inevitable destruction if not its formal experimentation. Following four doorstops of encyclopedic historical fiction from Stephenson, the equally voluminous *Anathem* is set in an alternate universe on the planet Arbre in the far future, but it is understood that this world's history mirrors the history of our own. Arbre's most distinct feature is that throughout the world there are large monasteries called concents. These concents were originally established after the end of the "Praxic Age" (when theory and practice were partners), an epoch that resembles our own present, at which point some global nuclear catastrophe occurred. After the destruction of civilization people decided to establish these concents for two reasons: to preserve both knowledge and the academic environment in which to pursue philosophy, math, and science unimpeded, and to organize these concents in such a manner that they would not be jeopardized by the vicissitudes of—nor would they affect—the outside world. By establishing strict rules regarding the technological advancement allowed to the monastic secular intellectuals inhabiting these concents, while severely restricting communication with the Praxic world, these monastery-archives were established to ensure that no matter what was happening outside, or "extramuros," they could preserve a degree of stability and continuity inside the archive itself while simultaneously restricting the knowledge and learning thought to have caused the ancient "Terrible Events" from practical application in the world. In other words, these archives are an attempt to prevent the remainderless archival destruction of total nuclear war by segregating the archive from the world in order to preserve the knowledge it contains. Further, each concent contains multiple barriers between different areas of the archive, called "maths," and these barriers are rigidly defined while being spatially porous at certain sanctioned times. Each section, or the math of each concent, is divided by a wall with a gate that opens every one, ten, hundred, or thousand years to communicate extramuros, the idea being that each successive level of the concent is more purely devoted to theory-as-such by being further removed from the outside world and the practical application of the knowledge stored in the archive.

Anathem's archive is thus located at a complex spatial and temporal frontier. Rather than residing in a deserted post-apocalyptic wasteland, as in *A Canticle for Leibowitz*, it is simultaneously at the center of the world and wholly outside of it. Likewise, for the avouts residing in the maths, the outside world constitutes a similar frontier, simultaneously outside their possible experience while limning their temporal and spatial existence inside, depending upon which section of the concent the avouts reside in. This radically shifts the "normative" post-apocalyptic narrative's representation of space as a liminal frontier (e.g., *Mad Max* [1979] or Cormac McCarthy's *The Road* [2006]). Rather than presenting post-apocalyptic space as simply something outside, distant, foreign, or other, something that the nuclear introduces through its transgressive destruction, in *Anathem* it also constitutes a center, ground, foundation, and core. Furthermore, rather than present contemporary civilization's collapse as temporally bounded, as eschatological or teleological, the civilization outside of *Anathem*'s concents, "[d]uring some eras [. . .] would grow and engulf our walls, and office workers in skyscrapers would gaze down on the tops of our bastions. At other times it would ebb and recede to a tiny fueling-station or gun emplacement at the river crossing"[36]; or as Erasmas—the protagonist of the novel—notes regarding the concent's honey production: "When there was an economy extramuros, we could sell the honey to burgers in the market stall before the Day Gate, and use the money to buy things that were difficult to make inside the concent. When conditions outside were post-apocalyptic, we could eat it."[37] These moments, along with much of the rest of the novel, display a concrete awareness that complex and inextricable relationships between technologies of information and destruction are further complicated and, in a way, reversed from traditional archival modes. The archive in *Anathem* becomes temporally teleological, and it is disaster that becomes a site of constant change, reversal, cyclicality, difference, and repetition. Indeed, in his off-handed observation of the inconsequentiality of whether the outside world was post-apocalyptic or not, Erasmas takes this archival reversal as a given.

As Erasmas's narrative moves forward it is revealed that a starship orbiting the planet contains humans from a different though parallel universe who have found a way to enter Arbre. Stephenson justifies the multiple universes within *Anathem* not only in terms of physics, but in terms of narrative. Each universe, or "worldtrack," has seen different historical narratives play out based on choices individuals have made, as well as attempts to violate the causal, linear nature of time itself: "This is how the universe protects herself—prevents violations of causality. If you attempt to do anything that would give you the power of violating the laws of cause-and-effect—to go back in time and kill your grandfather," one finds themselves "in a different and separate causal domain? How extraordinary! [. . .] One is shunted into an altogether different narrative [. . .] and thus causality is preserved." [38] In *Anathem*, not only notions of temporal and spatial liminality are upset by the crises of archivization—the relationship between the rise and fall of civilization and the archive's approach to knowledge preservation—like the fragmented, unreliable, and fabricated text of *Wilson*, narrative itself is subject to archival crisis. The archive in *Anathem* is able to project a world, and that ability is multiplied infinitely in Stephenson's invocation of alternate universes, conceiving space-time like some kind of hyper-Borgesian library. The universes of *Anathem*'s possible narratives are archives themselves, entries into the infinite, total set of all possible universes. In this way, this archive repeats, in a kind of vicious circle, its grounding in the archival crises everywhere punctuating the novel.

Situating the archive within a fluid, multiplicitous, synchronic, and recursive temporality is even more explicitly imagined in Charles Stross's "Palimpsest," which grounds itself in the understanding that, without major leaps in technology, leaps that appear at this point to be all but magic, humanity's future is indeed quite finite. As the short story repeatedly emphasizes, the universe cannot even indefinitely support itself. One can try to extend human existence as far and as long as the universe allows, but the extinction of humanity is inevitable no matter the extent of human technological prowess. The powers that be in the world of "Palimpsest," an organization known as Stasis, have one goal: to combat humanity's extinction for as long as it can. Having achieved the ability to send bodies and information forward and backward through time, Stasis seeds and reseeds the universe with humans over trillions of years. The narrative limits of "Palimpsest" correspond to the limits of the universe as such, much in the cosmic tradition of H.G. Wells, Jules Verne, or Olaf Stapledon.

Perched at the edge of a small crunch—in "Fimbulwinter: the winter at the end of the world, after the last fuel for the necrosun's accretion disk had been consumed, leaving Earth adrift in orbit around a cold black hole, billions of light-years from anything else"[39]—is an archive, "The Library at the End of Time,"[40] in which all of recorded human history resides. To accomplish their at times brutally preservationist goal, Stasis has constructed this immeasurably vast hyperarchive[41] to gather together all the data from the "fully ninety-six percent of humanity [that] lives in eras where the ubiquitous surveillance or personal life-logging technologies have made the recording of absolute history possible."[42] Collected during the collapse of whatever civilization produced it, the information is then sent forward to the archive, sent from a humanity who number "nearly twenty billion billion of us. We are not merely legion—we rival in our numbers the stars of the observable universe in the current epoch."[43] The Library then reconfigures and re-communicates the relevant information it contains to important points in the past, its teleological goal being to extend and preserve all historical timelines or "worldtracks" in which it exists. Archival preservation creates the history of archival preservation. This *ur*-library is not only a representation and agent of history, in a sense, it *is* history.

"Palimpsest" turns on the realization that the only way humanity could extend itself is temporally rather than spatially. The stars are too far to reach, the Kurzweilian fever of artificially extended posthuman life is not necessarily a desirable goal, and aliens, whether they exist or not, cannot reach us. The only hope for the survival of the species is through time, through the careful preservation, re-inscription, and reinterpretation of the archive so that it can communicate with the past and reconstruct a more desirable history. As such, any distinction between archival processes and history breaks down. The protagonist, Pierce, is a member of a select group of temporal policeman, known as *librarians*, who "are the eternal guardians of historicity, the arbiters of what really happened [. . .] sworn to serve our great cause—the total history of the human species."⁴⁴ This total history is, however, multiple. As is so often the case in time travel narratives, temporal paradoxes proliferate in Stross's world in the form of alternate timelines, but it is the librarian's job to shore up these paradoxes, to remove the moments of "palimpsest" when history threatens to become "an entire talmud of rewrites and commentaries [. . .] in a threatening tsunami of unhistory."⁴⁵ Consequently, "not only does the Library document all of recorded human history—and there is a *lot* of it [. . .]—it documents all the possible routes through history that end in the creation of the Final Library."⁴⁶ This Library is a kind of vast mystic writing pad; ninety-percent of it records events that did not actually "happen" and yet the entirety of the histories and "unhistories" it contains constitute the library's own meta-narrative. The hyperarchival processes of this Library gather together everything, even the many alternate histories—ultimately, *fictions*—that did not transpire, and yet nonetheless all of it directly shapes the Library's writing of (its own) *ur*-history.

Taken together, these three highly speculative fictions repeatedly demonstrate the systemic complexities available to the contemporary archival and eschatological imagination. Unlike *Earth Abides*, *On the Beach*, and *A Canticle for Leibowitz*, the archives of Mamet, Stephenson, and Stross are multiple and rhizomatic, fluidly situated in both space and time, and communicate history synchronically through multiple narrative modes that, though acknowledging various senses of an ending, articulate an eschatology that is quite different than either the utter destruction of the species through nuclear war or the complete deletion of the archive. Just as Stewart, Shute, and Miller rigorously channeled their speculations from the dominant anxieties and fantasies of the Cold War, *Wilson*, *Anathem*, and "Palimpsest" (like Lieberman's PCNAA), clearly emerge from the complex anxieties attending the global digital information culture. Rather than worry over the disappearance or destruction of information, each text, in its own way, frets about archival accumulation. Whether it is conceived as a tool to stave off the end of the world, or a system whose inevitable trajectory leads to destruction, the archive in each narrative is situated as a site of spatiotemporal crisis.

These tales of archival crisis dramatize and capture the relationship between today's increasingly totalizing and ubiquitous archival technologies, and the imaginative legacy of nuclear mass-destruction. The archive in each text is always-already a site of crisis, not having to await some global nuclear conflagration that would introduce a crisis into the text's narrative sweep. Further, it should be noted that "crisis" is derived from the Ancient Greek word *krisis*, meaning "decision," and its root verb *krinein*, meaning "to decide." (The Greek *krisis* could also mean "judgment" and "separation.") Drawing upon this etymology, these contemporary tales of archival crisis present moments of conflict not merely as windows onto material scenes of intense difficulty or danger, turning points or peripeteia; archival crises instead require some kind of decision, and this decision is an archival decision. To put it another way, we might look at the tale of archival crisis in terms of material emergence—i.e., a mass of chaotic material organizing into a higher level of order seemingly spontaneously in order to respond to the crisis. In this fashion, each of these twenty-first century tales of archival crisis should be understood as engaging (hyperbolically) with how digital archival systems write and rewrite themselves, creating social and technological nodes that respond to and dictate history. Each text attributes certain properties to the archive in which different historical, political, and discursive forces emerge, often in the absence of human intervention. Consequently, not only can apocalyptic levels of destruction emerge from the hypothetical archives imagined in these texts, but so can another fantasmatic limit: the creation of a world.

Conclusion: The Archive to Come; or, the Task(s) of Nuclear Criticism at the Present Time

In light of the emergence of these tales of archival crisis of the second nuclear age, it should now be clear that a post-Cold War nuclear critical practice would greatly benefit from returning to a moment from Derrida in which he should be read as slightly revising his earlier hypothesis of the total and remainderless destruction of the archive. In a talk given prior to a discussion with book historian Roger Chartier and philosopher Bernard Stiegler in 1997, Derrida proposed another archival limit and hypothesis that deserves to be quoted at length:

> Now what is happening today [. . .] of the book's to-come, still as the book, is *on the one hand*, beyond the closure of the book, the disruption, the dislocation, the disjunction, the dissemination with no possible gathering, the irreversible dispersion of this total codex [. . .]; but simultaneously, *on the other hand*, a constant reinvestment in the book project, in the book of the world or the world book, in the absolute book. [. . .] It re-creates the temptation that is figured by the World Wide Web as the ubiquitous Book finally reconstituted, the book of God, the great book of Nature, or the World Book finally achieved in its onto-theological dream, even though what it does is repeat the end of the book as to-come.

There are two fantasmatic limits of the book to come, two extreme, final, eschatic figures of the end of the book, the end as death, or the end as *telos* or achievement. We must take seriously these two fantasies; what's more they are what makes [sic] writing and reading happen. [. . .] But we should also perhaps wake up to the necessity that goes along with these fantasies.[47]

The necessity that goes along with these fantasies, which Derrida develops as four interminable "vanishing points," is what the post-Cold War nuclear archive gives to thought: the archiving of the world, the hypothesis of a total and remainderless documentation or creation of a world. If the first nuclear age, according to Derrida, introduced total archival destruction to thought, the second nuclear age has given it (or made it remember) total archival creation. And these are equally threatening. "The truth of the book, if I may put it like that, at any rate its *necessity*, resists—and dictates to us (this is also the seriousness of a 'must') that we should resist both these fantasies, which are only the flipside of each other."[48] As he once suggested that "[t]he hypothesis of this total destruction watches over deconstruction,"[49] today the hypothesis of total capture must watch over a revived Nuclear Criticism.

Regardless of their obvious differences, Mamet, Stross, and Stephenson's texts are born out of and respond to the ubiquitous archival surveillance of the present age, the urge—as Microsoft's *MyLifeBits* project[50] and the basic logic of Google demonstrate—to capture, record, document, and store *everything*. Obviously this is a fantasmatic limit, but it is a fantasy we should take seriously. We must be aware of the relative inhumanity that drives and results from the hyperarchival impulse. As Alexander Galloway and philosopher Eugene Thacker convincingly argue, when we start exploring network science and its relationship to archival processes, something emerges which "is that of *network being*, a *Dasein* specific to network phenomena."[51] Today, as Lieberman's PCNAA limply gestures toward, yesterday's archives have grown out of our control to understand, map, or destroy them. Like the force Henry Adams famously symbolized with the Dynamo, and most assuredly the cultural effects created by perceptions of the nuclear bomb, the archive both is, and should be considered a symbol for, a material and historical force, one with great creative as well as destructive potential.

In this fashion, we would do well to read the fate of the nuclear referent after the Cold War as a *dispersal* rather than a disappearance. One could argue that the U.S. cultural sense of an ending, its ability to imagine itself at the brink of some disaster, actually increased and diversified during the 1990s and 2000s.[52] The sheer proliferation of apocalyptic fantasies in the popular imagination, whether they imagine ecological, religious, viral, natural, or technological disaster, considered along with the frequent real world catastrophes that marked the past two decades, are existentially overwhelming if dwelt upon for too long. To quote Fredric Jameson, revising an earlier statement, "Someone once said that it is easier to imagine the end of the world than to imagine the end of capitalism. We can now revise that and witness the attempt to imagine capitalism by way of imagining the end of the world."[53]

Along with this contemporary archive of disaster, we should also emphasize the hyperarchival nature of emerging texts themselves. With the increasing space made available by improvements in information technology, texts are now being created that are simply massive, unthinkably large in terms of most recording technologies' capacity during the twentieth century. Take, for example, Richard Grossman's forthcoming *Breeze Avenue* (2014), a three-million page mega-text, that according to Grossman's website will "[manifest] itself most significantly as a set of 5,000 volumes, each containing 600 pages that are installed in a reading room. Although available in digital formats, in its massive printed version it stands as a bulwark against predigested and inane communication and as symbol and dirge for the global extinction of culture now in progress."[54] The term "encyclopedic" does no justice to the size of such a text, which will be installed literally as an archive. The contemporary mega-text, and the mega-narratives now proliferating, both in video games and more "traditional" cultural forms like television and the novel,[55] materialize one limit of Derrida's absolute-book-to-come through their sheer unreadable size, while also constantly drawing attention to their own archive fever, their own (futile) resistance to decay, destruction, disappearance, and rewriting.

So we should ask, given the atmosphere of this global network of doom, both real and imagined, what role should Nuclear Criticism play going forward? To imagine the task of Nuclear Criticism at the present time we should begin by remembering that in certain ways it perhaps became unnecessary in the immediate wake of the Cold War. Whether this statement is read as overly modest or hubristically grandiose, certain real effects have been achieved by the continued critique of nuclear power, not the least of which is the simple fact that there has been no nuclear war since Hiroshima. The Nuclear Critic is now in the strange position in which her object of study has seemed to disappear and this is something to be celebrated. And yet the global imaginary churns out post-apocalyptic images on a daily basis. As some have convincingly suggested, perhaps a risk criticism or a disaster criticism would be a more appropriate endeavor than Nuclear Criticism at the moment.[56] To my mind, however, it is the nuclear that continues to define the background of the contemporary eschatological imagination, remaining in a variety of forms as a kind of nuclear dust. The sense of an ending that the thermonuclear bomb introduced to the species (and the archive) has become multiple (revealing in hindsight that it always was). From ecological, economic, and epidemiological catastrophes, to the emerging informational, networked, and posthuman eschatologies, the nuclear bomb's continued and diverse representation and expression across a range of media is frankly staggering, disturbing, and fascinating. This alone should continue to pose a rich and vital task for the continued practice of Nuclear Criticism.

Further, there is a marked anxiety in literature, criticism, publishing, and the university over the effects of increased digital archivization, an anxiety which ranges from fears about the death of print, the end of libraries, and the foreclosure of the humanities, to fears that extend to the posthuman extinction (or transformation) of the species. These anxieties reveal that, even if overtly specific representations of nuclear war have largely disappeared from the U.S. imaginary, there remains a marked dialectical and structural tension between the representation of technologies of destruction and accumulation, between information and nuclear technologies, between crisis and possibility. Revisiting Nuclear Criticism, especially as I have framed it here, can enhance the methods and perspectives from which critics can continue the very necessary task of accounting for cultural production in the information age, a criticism that is capable of rigorously attending to the materiality of texts, the technologies through which information flows, accumulates, and disappears, and the networked, distributed nature of the contemporary apocalyptic imagination. Even more so, Nuclear Criticism may ultimately prove to offer a mode of confronting our own sense of the textual, disciplinary, and discursive crises that characterize so many aspects of the world today. By revisiting Nuclear Criticism we might find some quite useful tools to account for the persistent rhetoric of crisis that the network society appears incapable of escaping and radically change the conversation.

Notes

[1] Jorge Luis Borges, "The Book of Sand" (1975), in *Collected Fictions*, trans. Andrew Hurley (New York: Penguin, 1998), 483.
[2] Declan McCullagh, "Senators Propose Granting President Emergency Internet Power," *CNET*, 10 June 2010, http://news.cnet.com/8301-13578_3-20007418-38.html. As McCullagh points out, "The idea of an Internet 'kill switch' that the president could flip is not new. A draft Senate proposal that *CNET* obtained in August [2009] allowed the White House to 'declare a cybersecurity emergency,' and another from Sens. Jay Rockefeller (D-W.V.) and Olympia Snowe (R-Maine) would have explicitly given the government the power to 'order the disconnection' of certain networks or Web sites." The full text of the PCNAA is available at http://www.opencongress.org/bill/111-s3480/text.
[3] See Chloe Albanesius, "Lieberman Backs Away from 'Internet Kill Switch,'" *PcMag.com*, 21 June 2010, http://www.pcmag.com/article2/0,2817,2365393,00.asp.
[4] Fredric Jameson, *Postmodernism; or, the Cultural Logic of Late Capitalism* (Durham: Duke University Press, 1991), 38. For a discussion of the nuclear sublime, see Frances Ferguson, "The Nuclear Sublime," *diacritics* 14, no. 2 (Summer 1984): 4-10.
[5] On these latter points, see Ulrich Beck, *World at Risk* (2007), trans. Ciaran Cronin (Malden, MA: Polity Press, 2009).
[6] Alexander R. Galloway, *Protocol: How Control Exists After Decentralization* (Cambridge, MA: The MIT Press, 2004), 29.
[7] Ibid., emphases mine. It should be further noted that U.S. nuclear policy *itself* has adapted to the new paradigm of what Galloway calls the distributed network. Jonathan Schell perceptively analyzes this structural and systemic change in policy in *The Seventh Decade*, noting Donald "Rumsfeld's most famous articulation of this new strategic uncertainty principle was that you must plan not only for the 'known unknowns' but for the 'unknown unknowns.' In the last analysis, the target of the nuclear arsenal became history and whatever it might produce—not a foe but a tense, the future itself." In Jonathan Schell, *The Seventh Decade: The New Shape of Nuclear Danger* (New York: Metropolitan Books, 2007), 121.
[8] It is also interesting to note that the first particularly American archival theory, articulated in T.R. Schellenberg's *Modern Archive: Principles and Techniques* (Chicago: Society of American Archivists, 1956), emerged in the aftermath of World War II to account for the massive number of government documents that were produced during the war. For a brief history of archival theory, see John Ridener, *From Polders to Postmodernism: A Concise History of Archival Theory* (Duluth, MN: Litwin Books, 2009).
[9] This is one of the tenets of Manuel De Landa's crucial text on the subject, *War in the Age of Intelligent Machines* (New York: Zone Books, 1991).
[10] For excellent recent nuclear critical studies of Cold War anxieties in American literature, see Daniel Grausam, *On Endings: American Postmodern Fiction and the Cold War* (Charlottesville, VA: University of Virginia Press, 2011) and Daniel Cordle, *States of Suspense: The Nuclear Age, Postmodernism and United States*

Prose (Manchester, UK: Manchester University Press, 2008). Cordle is also a contributor to this volume; see chapter 11, "Legacy Waste: Cultural Fallout and Nuclear Criticism in the Twenty-First Century." For a comprehensive and nuanced overview of Nuclear Criticism as it was practiced between 1984 and 1993, see Ken Ruthven, *Nuclear Criticism* (Victoria, AU: Melbourne University Press, 1993).

[11] Jacques Derrida, "No Apocalypse, Not Now (Full Speed Ahead, Seven Missiles, Seven Missives)" (1984), trans. Catherine Porter and Philip Lewis, in *Psyche: Inventions of the Other*, Vol. 1, eds. Peggy Kamuf and Elizabeth Rottenberg (Stanford: Stanford University Press, 2007), 400. For an extended meditation on the relationship between the nuclear bomb and the literary archive, see Peter Schwenger, *Letter Bomb: Nuclear Holocaust and the Exploding Word* (Baltimore: The Johns Hopkins University Press, 1992).

[12] It is important to remember that Derrida gives just such (destructive) agency to archives ten years after the Cornell conference on Nuclear Criticism in *Archive Fever: A Freudian Impression* (1995), trans. Eric Prenowitz (Chicago: The University of Chicago Press, 1996).

[13] Fernando Báez, *A Universal History of the Destruction of Books: From Ancient Sumer to Modern Iraq* (2004), trans. Alfred MacAdam (New York: Atlas & Co., 2008), 6-7.

[14] Derrida, *Psyche*, 400.

[15] Saint-Amour is also a contributor to this collection. See chapter 3, "Queer Temporalities of the Nuclear Condition."

[16] His reading of modernism and Joyce are of particular note, and I think much can be gained from continuing his project of reading the pre-Hiroshima nuclear archive. See Paul K. Saint-Amour, "Bombing and the Symptom: Traumatic Earliness and the Nuclear Uncanny," *diacritics* 30, no. 4 (Winter 2000): 59-82; "Over Assemblage: *Ulysses* and the *Boîte-en-valise* From Above" in *Cultural Studies of James Joyce*, ed. R. Brandon Kershner, *European Joyce Studies* 15, no. 1 (New York: Rodopoi, 2003), 21-58; and "Air War Prophecy and Interwar Modernism," *Comparative Literature Studies* 42, no. 2 (2005): 130-61.

[17] Walter M. Miller, Jr., *A Canticle for Leibowitz* (New York: Harper and Row, 1959), 11.

[18] Similar post-nuclear misreadings occur in Russell Hoban's *Riddley Walker* (1980), expanded ed. (Bloomington, IN: Indiana University Press, 1988) and Will Self's *The Book of Dave: A Revelation of the Recent Past and the Distant Future* (New York: Bloomsbury, 2006).

[19] For a brief history of the literary representation of the destruction of books and archives, see Báez, 188-198.

[20] Carolyn Steedman, *Dust: The Archive and Cultural History* (New Brunswick: Rutgers University Press, 2002), 164, emphases in original.

[21] Derrida calls this the "absolute *epochē*" of the nuclear age, *Psyche*, 401.

[22] James Berger, *After the End: Representations of Post-Apocalypse* (Minneapolis: University of Minnesota Press, 1999), 5-6, emphases in original.

[23] Ibid., xiii.

[24] This eschatologically anticipatory archival project is not limited to literary fiction. For example, in 1989 it was reported that the Internal Revenue Service, an immense archival institution, has a plan in place to continue to collect taxes after a thermonuclear war. See "Nuclear War Plan By I.R.S.," *The New York Times,* 28 May 1989, http://www.nytimes.com/1989/03/28/business/nuclear-war-plan-by-irs.html. The U.S. National Gallery also has a "special archive" whose goal is the preservation of works in the wake of a national catastrophe. See Ned Martel, "Curator Andrew Robison Decides What Goes into National Gallery's Emergency Box," *The Washington Post,* 14 August 2011, http://www.washingtonpost.com/politics/curator-andrew-robison-decides-what-goes-into-national-gallerys-emergency -box/2011/08/08/gIQAUTVsFJ_story.html.

[25] Walter Benjamin, "Theses on the Philosophy of History" (1940), in *Illuminations: Essays and Reflections,* trans. Harry Zohn, ed. Hannah Arendt (New York: Schocken Books, 1968), 256.

[26] Along with Peter Schwenger, for the more explicitly deconstructionist Nuclear Criticism, see Avital Ronell, "Starting From Scratch: Mastermix," *Socialist Review* 18, no. 2 (April-June, 1988): 73-86; Richard Klein, "The Future of Nuclear Criticism," *Yale French Studies* 77 (1990): 76-100; and Roger Luckhurst, "Nuclear Criticism: Anachronism and Anachorism," *diacritics* 23, no. 2 (Summer, 1993): 89-97. Critics such as J. Fisher Solomon, William J. Scheick, and others, though clearly indebted to Derrida, took issue with Nuclear Criticism's emphasis on the textuality of the nuclear referent, wanting instead to practice a more ethical Nuclear Criticism, one that stressed the reality, rather than poststructural textuality, of nuclear weaponry and the possibility of its deployment. See J. Fisher Solomon, "Possible Circumstances, Potential Worlds: History, Futurity, and the 'Nuclear Referent,'" *Papers in Language and Literature* 26, no. 1 (1990): 60-72 and *Discourse and Reference in the Nuclear Age* (Norman, OK: University of Oklahoma Press, 1988); William J. Scheick, "Nuclear Criticism: An Introduction" and "Atomizing a Postage Stamp (1955)," in *Papers in Language and Literature* 26, no. 1 (1990): 3-12, 182-5; and Christopher Norris, *Uncritical Theory: Postmodernism, Intellectuals, and the Gulf War* (Amherst: The University of Massachusetts Press, 1992).

[27] Derrida, *Psyche*, 393.

[28] In her recent study, *Too Much to Know: Managing Scholarly Information before the Modern Age* (New Haven, CT: Yale University Press, 2010), Ann M. Blair demonstrates that the phenomenon of encountering and attempting to sort through too much information—an encounter with the informational sublime—is not a recent phenomenon, and has a long history that can be traced back to antiquity.

[29] Derrida, *Psyche*, 400.

[30] It should also be noted that David Mamet has had a much publicized rift with his previously held liberal-democratic viewpoints, leading him to embrace neoconservatism. See David Mamet, *The Secret Knowledge: On the Dismantling of American Culture* (New York: Sentinel, 2011) and Jordan Michael Smith, "Was David Mamet Ever Really a Liberal Anyway?" *The Awl,* 20 June 2011, http://www.theawl.com/2011/06/was-david-mamet-ever-really-a-liberal-anyway.

[31] One might also suggest that the novel does this to an offensive degree, which may account for the roundly negative reviews it has received. See John Fortune, "Footnotes and Fancy Free: The Question is This: Is David Mamet's *Wilson* a Work of Genius or a Vast Pile of . . . ?" *The Observer*, 27 February 2000, http://www.guardian.co.uk/books/2000/feb/27/fiction.davidmamet.

[32] David Mamet, *Wilson: A Consideration of the Sources, Containing the original Notes, Errata, Commentary, and the Preface to the Second Edition* (Woodstock: The Overlook Press, 2000), back cover.

[33] Ibid., xiii.

[34] Ibid., 20-21.

[35] Ibid., xiv.

[36] Neal Stephenson, *Anathem* (New York: William Morrow, 2008), 60.

[37] Ibid., 15.

[38] Ibid., 712.

[39] Charles Stross, "Palimpsest," in *Wireless* (New York: Ace, 2009), 282.

[40] Ibid., 272.

[41] "Hyperarchive" is a term that I think nicely designates an archive whose goal, whether stated or not, can be seen in an attempt to gather together as many documents and texts as it can, regardless of content.

[42] Stross, 250.

[43] Ibid., 249.

[44] Ibid., 289.

[45] Ibid., 301.

[46] Ibid., 287.

[47] Jacques Derrida, "The Book to Come," in *Paper Machine*, trans. Rachel Bowlby (Stanford: Stanford University Press, 2005), 15. This essay draws heavily upon Maurice Blanchot's *The Book to Come* (1959), trans. Charlotte Mandell (Stanford: Stanford University Press, 2003), and its reading of Stéphane Mallarmé's "*Un coup de dés jamais n'abolira le hasard*" (1897).

[48] Derrida, *Paper Machine*, 17, emphasis in original.

[49] Derrida, *Psyche*, 400.

[50] On Gordon Bell's project to document and record his entire life, see Alec Wilkinson, "Remember This?" *The New Yorker* 83, no. 14 (28 May 2007): 38-46, http://www.newyorker.com/reporting/2007/05/28/070528fa_fact_wilkinson?currentPage=all.

[51] Alexander R. Galloway and Eugene Thacker, *The Exploit: A Theory of Networks* (Minneapolis: University of Minnesota Press, 2007), 118.

[52] From James Cameron's 1991 film *Terminator 2: Judgment Day*, in which NORAD became self-aware and began to eradicate humanity, to the disaster spectacles of Roland Emmerich (*Independence Day* [1996], *The Day After Tomorrow* [2004], and *2012* [2009]), to the revivification of the zombie genre (see Sarah Juliet Lauro and Karen Embry, "A Zombie Manifesto: The Nonhuman Condition in the Era of Advanced Capitalism," *boundary 2* 39, no. 1 [Spring 2008]: 85-108), to the ecological disasters of Kim Stanley Robinson's *Science in the Capital Trilogy* (2004-2007) or Margaret Atwood's *Oryx and Crake* (2004)

and *The Year of the Flood* (2009), there has been no shortage of diverse, multiplicitous, and seemingly ubiquitous disaster narratives during the second nuclear age. Indeed, to begin to account for the full range of such narratives since the end of the Cold War in film, literature, television, and elsewhere would require a collection of its own. For two recent studies of twentieth and twenty-first century disaster see Peter Y. Paik, *From Utopia To Apocalypse: Science Fiction and the Politics of Catastrophe* (Minneapolis: University of Minnesota Press, 2010) and Evan Calder Williams, *Combined and Uneven Apocalypse* (Winchester, UK: Zero Books, 2011).

[53] Fredric Jameson, "Future City" (2003), in *The Ideologies of Theory* (New York: Verso, 2008), 573.

[54] See *RichardGrossman.com*, http://www.richardgrossman.com/breeze_avenue/.

[55] For some interesting beginning efforts to delineate the mega-narrative as a field of study, see Pat Harrigan and Noah Wardrip-Fruin eds., *Third Person: Authoring and Exploring Vast Narratives* (Cambridge, MA: The MIT Press, 2009).

[56] In Molly Wallace's "Will the Apocalypse Have Been Now? Literary Criticism in an Age of Global Risk," she argues that that Nuclear Criticism could learn much from ecological criticism and vice versa, and together could constitute something like a *risk criticism* for the second nuclear age. In *Criticism, Crisis, and Contemporary Narrative: Textual Horizons in an Age of Global Risk*, ed. Paul Crosthwaite (New York: Routledge, 2011), 15-30.

Chapter Five

Cut to Black: Nuclear Criticism in a Post-September 11th America

Joseph Dewey

> We must get it out of our heads that this is a doomed time…Things are grim enough without these shivery games. People frightening one another—a poor sort of moral exercise. The praise of suffering takes us in the wrong direction…Why not say rather that people of powerful imagination, given to dreaming deeply, to raising marvelous fictions, turn to suffering sometimes to cut into their bliss, like people pinch themselves to feel awake.
> —Saul Bellow, *Herzog*

> These are the days after. Everything now is measured by after.
> —Don DeLillo, *Falling Man*

What the -----?

It was the collective gasp of the HBO Nation. June 10, 2007, an otherwise nondescript late Sunday evening when, across seven time zones, more than eleven million *Sopranos* fans, anticipating the final scene of the groundbreaking mob drama, group-gasped as their television screens suddenly, inexplicably cut to black. Not the gray clatter-wash of static, not the friendly blue screen that reassured that technical difficulties were being worked out—just a forbidding black. Censored? After six years of passing along scenes of graphic brutality and serial promiscuity, now, in the last few minutes, the censor was suddenly going to intrude? Couldn't be. In unchoreographed choral panic, viewers vigorously shook their remote controls to jar the batteries just enough to restore the signal for the few minutes left in the series; others scatter-flipped to quick-check whether the blackout was systemic or restricted to HBO. But the cut to black was absolute only to HBO. Unbelievable. At the pivotal moment when Tony Soprano, who had for six charmed seasons contained, controlled, directed, and trafficked in death all the while eluding a platoon of dogged federal prosecutors, the entangling webs of FBI sting operations, disgruntled underworld associates, and jackal-like rival gangsters, was himself poised to succumb to the cannibalism of his own business model, poised for his karmic come-uppance, a stroke as inevitable as it would be violent, as all aficionados of the American gangster culture would acknowledge. Now facing imminent federal indictment, ratted out by a weasely informant, the once-efficient operations of his own crew collapsing as much from hubris as carelessness, Tony was poised to embrace what for more than eighty episodes he had so defiantly avoided: the blood-ending that narrative has always reserved for particularly charismatic villains. Ironically, after six years of cheering while Tony flaunted the law with Rabelaisian gusto, eleven million fans now sanctimoniously anticipated that justice would at last be served, and, fittingly, the bloody kind of street justice that Tony himself had for so long administered from the shadowy room of Satriale's Pork Shop.

The episode had certainly set up such anticipation. Before the cable glitch, viewers had been ushered with gathering momentum toward a tightly orchestrated (and heavily symbolic) Last Supper: Tony sharing a plate of onion rings with his long-suffering wife Carmela and his troubled son A.J., the three sitting in an upholstered booth in a clean, well-lighted New Jersey diner. Leading up to the closing minutes, we had felt the heavy presence of a forbidding tension. Viewers had been prepped for cataclysm. Despite a shaky truce, Tony's men had moved on a rival boss, shooting him dead in a gas station in front his family, setting up the logic of reprisal. And there was A. J.—whose spiffy SUV had exploded when he had parked it over a pile of leaves that had inexplicably caught fire—improbably muttering lines from Yeats' apocalyptic "Second Coming." There was Tony's inexplicable (and uncharacteristic) fondness for a mangy stray cat, symbolizing death, we were confident, specifically Tony's. The last scene itself was freighted in overtly suggestive symbolism, particularly the clusterings of threes (three Cub Scouts in a nearby booth; three Sopranos at the table; three light fixtures on the back wall framing Tony; three creamers on a customer's absurdly large coffee cup; three rings of the diner's entrance bell; and, outside, his daughter's three tries at parallel parking). Then there were the gratuitous shots of an unfamiliar man who comes in and sits at the counter, wearing a jacket bearing the logo of Members Only (the Mafia?), who suddenly gets up and heads to the bathroom. Keen to the obvious allusion to the classic shootout scene in *The Godfather* when the reluctant gangster-in-waiting Michael Corleone earns his bones when he emerges from the bathroom in Louis's

Restaurant, gun in hand, to dispatch the rival crime boss and the corrupt police captain responsible for the attempted killing of his father, fans edged toward the screen, bracing for the approaching blood bath. Even as Tony absently selects Journey's "Don't Stop Believin'" from the diner jukebox, even as the song's familiar piano opening blares, even as the now-ironically nonplussed Tony calmly raises a fried onion ring to his mouth, even as the little bell over the diner entrance jangles, signaling presumably his daughter's tardy arrival, Tony, with that familiar crooked smile, looks up and then...

Cut to black, ten lingering seconds of a formidable absolute black.

What the ----?

To a still-traumatized post-September 11th American culture hungry for certainty and uninterested in the crazy subtlety of paradox, the David Chase ending was a challenge to process; ready to engage the heavy theatrics of catastrophe, viewers were left uneasily suspended between and among resolutions. It is certainly not to compare the last episode of a television series—even a landmark achievement such as *The Sopranos*—to the apocalyptic end of a nation or an ecosystem, but rather to suggest that the rich energy and moral vigor of Nuclear Criticism, summarily shunted to the margins in the faux-euphoria following the collapse of the Soviet Empire in the early 1990s, might help explicate the implications of the David Chase script; indeed that dramatic cut to black manifests the very argument that could, indeed should, invite Nuclear Criticism if Nuclear Criticism might be wrested free of its assumed anchorage in the melodramatic anxieties of the High Noon rhetoric of the Cold War, now nearly three decades in the past. Examining Chase's "Made in America" episode—and constellating it with other cultural artifacts of post-September 11th America—is to suggest that the foundation concepts of Nuclear Criticism animate far more than the half-century response to the broad and terrifying implications of the bombings of Hiroshima and Nagasaki. The bold agenda of Nuclear Criticism is neither that parochial nor that precious. Rather the audacious Chase script can be seen as part of a much wider genre of narrative that embraces not merely the usual suspects in the speculative fiction canon or the rhetoric extravagances of Biblical eschatological writings but cutting-edge Western texts that, since Defoe's *The Journal of the Plague Year* (1722), have worked to reclaim the open mystery of tomorrow against the formidable reality of approaching cataclysm brought about by an oppressive industrial-scientific-military-political construct bent on diminishing that tonic wonder.

Consider those exhilarating/aggravating ten seconds when the HBO screen cut to black. Hungry for the simplification of imminent closure, eager for the tidy resolution that is the largesse of narrative, viewers were denied such privileged position and dropped rudely into the wide animated energy field of a difficult and dense "perhaps," pitched into the formidable ironic matrix of "possibility." Was Tony dead? Was Tony alive? Was his entire family killed? Why had A.J. walked in with the man in the Members Only jacket—was the goth son finally striking back at the father? Was the tardy Meadow in fact pulling her own Michael Corleone power play? Or was this entire Last Supper actually one of Chase's signature dream sequences? To all of the above, a bemused Chase would enigmatically nod—he hints as much when Tony makes his jukebox selection: the flipside is Journey's "Any Way You Want It." Fans were left exposed, vulnerable within a generous extra-textual space of absolute mystery—a reminder, as works in the field of Nuclear Criticism argue, that clear sight and insight are never the same. Fans were compelled to recalibrate their narrative sextants, left in the crazy inkscape where the American imagination has thrived since Melville, to press full throttle into the fog. Not surprisingly the non-ending ending generated a firestorm of criticism among the show's faithful in the blogosphere. Angry, they dismissed the show's non-closing as frustrating, baffling, and even cowardly. The "Made in America" episode exposes the tensions between the seductive patterning and coaxing logic typical of the entirely symbolic, protective safe house of narrative and the wide open play of possibilities typical of texts of Nuclear Criticism. This tension, in turn, has long rested at the heart not only of America's decade-long response to the forbidding implications of September 11th but in its long history of responding to the perception that this generation will surely be the last. This thought process dates back in the American experience to the first generation Puritans who saw the rugged expanse of coastal wilderness only as backdrop with dimensions sufficient for the imminent descent of the new earth. As such, "Made in America" is a pitch-perfect text, a catapult act of audacious insight, that affirms the (re)animation and continuing relevancy of Nuclear Criticism in a post-September 11th culture, a culture now on permanent color-coded alert, a culture rudely returned to the anxieties of anticipating cataclysm.

II

Because some men aren't looking for anything logical, like money. They can't be bought, bullied, reasoned or negotiated with. Some men just want to watch the world burn.
—*The Dark Knight*, Christopher Nolan (2008)

Nuclear anxiety, which initially generated and then sustained Nuclear Criticism, today suffers from an image problem brought on by its own branding. The adjective compels its narrow positioning as an expression of an argument now widely considered obsolete, an exercise in a kind of cultural nostalgia for the era of hula hoops and drive-ins; and the noun, so forbidding and so extreme, suggests hyper-apprehensive texts, dark and pessimistic narratives, buzzkills from a gathering of obdurate, whiney Chicken Littles. That image problem itself is curious—after all, when we interrogate the position of Nuclear Criticism *after* the Cold War era, we are still essentially talking about what we talked about when we interrogated the position of Nuclear Criticism *during* the Cold War era. Indeed, whatever the giddy euphoria generated as we watched those sledgehammers fragment the Berlin Wall, we have never been out of the Nuclear Age, have lived now for nearly eighty years under the formidable arc of gravity's rainbow—the forbidding reality that humanity has never assembled a weapons system that it did not ultimately unleash. The wide-range of post-Hiroshima texts initially validated within the burgeoning field of Nuclear Criticism—novels, television shows, film, plays, cultural histories, scientific data, investigative journalism, memoirs—centered not surprisingly on exploring the implications of the era's tense political realities, scenarios ripped from headlines; the alarming stats on nuclear weaponry capabilities; the delicate brinksmanship of diplomatic and military posturing; the sobering scientific data on the dimensions of the devastation that would be unleashed should the nuclear stockpiles, now quiet and harmless, go singing free of their silos.

These first generation texts of nuclear anxiety were rare unicorn-works. They fit conveniently into neither of the then-dominant branches of narrative imagination: they were neither the dazzling gamesmanship of trendy, writerly texts that so tirelessly foregrounded formal experiments and language theory nor were they expressions of the taunt psychological realism that compassionately chronicled the bruising busyness of small, strangled lives. Rather, these first-generation texts of Nuclear Criticism— among them, Vonnegut's *Cat's Cradle* (1963); Percy's *Love in the Ruins* (1971); Pynchon's *The Crying of Lot 49* (1966); DeLillo's *End Zone* (1972); Malamud's *God Grace* (1982); Bellow's *Herzog* (1964); Walter Miller's *A Canticle for Leibowitz* (1960)—with a singular gravitas and uncompromising Cassandra-like vision, went about the difficult business that since John the Divine chained to the rocks on Patmos has been the compelling protocol of texts of cultural (though not always specifically nuclear) anxiety: nothing less than the construction of a viable hope for a culture suddenly, dramatically denied the logic of continuity. That first generation of Nuclear Criticism, inevitably, anatomized texts written within the heavy shadow of the Hiroshima mushroom cloud, texts that dared to think about the Unthinkable, to bring to its culture the stark realities of what-if, defined not by the extravagant fantasies of the alternate universes of speculative fictions but rather by the hard data of Defense Department projections of first-strike scenarios and acceptable civilian losses. In Nuclear Criticism was a topic—nothing less than species self-extermination—too big, too pressing, too real for the cloying academic play of theoretical dissection.

Apocalypse was suddenly not a harmless metaphor—rather a viable strategy, being war-gamed in classified binders in the securest reaches of the Pentagon. In bringing together such a variety of disciplines—journalism, history, sociology, theology, psychology, cultural studies, film studies, environmental science, media studies, political rhetoric, as well as fiction and poetry and drama—Nuclear Criticism promised a new, ambitious school of thought, designed not for the classroom but for the community, ministering to a fractured, anxious community locked in crisis, and to extend to that community the tonic consolation of hope in a dark and difficult time by restoring that troubled community to the difficult privilege of uncertainty. These courageous texts argued that narrative could not fix the endangered world, could not disentangle the wonderland logic of Mutual Assured Destruction, could not protect the vulnerable; instead narrative could minister to those made vulnerable by offering case studies of those who must engage the uncertain improvisation of the everyday against and amid the certainty of imminent endings.

It is odd then that the adjective most often applied to Nuclear Criticism is "short-lived." Given the sheer kinetic dimension of nuclear anxiety generated during the height of the reckless High Noon rhetoric of the early Reagan Era, it is perhaps not entirely surprising that with the heavily symbolic (and irresistibly allegorical) tearing down of the Berlin Wall in 1989, and the subsequent exposure of the long-feared Soviet menace as little more than a paper bear, the visionary element of foundation critical works in the field was swiftly ushered to the margins—despite the obvious fact that in the quarter-century since the fall of the Soviet Empire, nothing has significantly altered the dimension of the nuclear arsenal that still waits in silo launchers for its inevitable deployment. Indeed, the temper of Ken Ruthven's elegant *Nuclear Criticism* (1993), often cited as the last, best word in the genre, sounded a rueful note of valedictory in its closing pages, reminding us that the argument of Nuclear Criticism, in the wake of the (so-called?) New World

Order, could still be applied profitably to the growing alarums raised by the international environmental movement and by our deep-seated misgivings about nuclear power industry itself. Indeed Ruthven, like some lonely John the Baptist in the wilderness, called for the tenets of Nuclear Criticism to frame the argument within the humanities into the new century.

It turned out not to be such an easy sell. During the 1990s, primary texts, fiction and nonfiction, appeared no longer drawn to the premise of nuclear anxiety—and the tone of much scholarly work on the subject, in turn, reworked nuclear anxiety into a kind of wistful Happy Days nostalgia. Energized by the engaging youth and vigor of the Clinton White House, invested in the decade's giddy economic gladtimes, and happily distracted by the loopy decade-long wacky investigations into assorted Clinton scandals, we found anxiety—nuclear or otherwise—refreshingly ironic. Given that decade's gleeful surrender to the centripetal hypnosis of cyber-realities as cable television exponentially expanded its reach; given its willing purchase, encouraged by the endless patter of both radio call-in shows and television talk shows, into the dynamic of the happy inevitability of quick-fix solutions; given its contented interment within the engaging soft prisonhouse of entertainment technologies, specifically a generation of compelling video games that rendered the drama of conflict harmless and even fetching; given its deployment of American military forces in exotic and decidedly contained engagements; given its rendering into harmless parody a succession of earnest latter-day prophets each with a timetable for Creation's approaching end; given its jaded cynicism over the viability of law and order and the process of justice itself in the wake of both the O. J. Simpson trial and the subsequent Simi Valley trial; and given its wide embrace of the template of a generous and gentle Christian God (promoted by the tentacled reach of televangelists and the vociferous advocates of the now-entrenched political Right), the 1990s offered a provocative and suasive hiatus from anxiety, a disengaged and complacent culture uninterested in confronting imminent peril.

Certainly there were disquieting mutterings—discontent over the inconvenient truth of the slow-burn of global warming and the consequent quiet destruction of the greenhouse effect; concerns over radical terrorists (most notably the Oklahoma City bombing); the carnival endgame of the 2000 presidential elections; and supremely a Y2K clickover computer glitch that might cripple entire economies, compromise military security systems, and even bring jets raining down. But the decade was in no mood for anxiety, perhaps taking its cue from the thematic argument of the decade's signature cinema event, James Cameron's upbeat (and visually immersive) epic recreation of the *Titanic*: large-scale catastrophe makes for engaging (and harmless) theater.

III

> We wanted to find America through the gasps of snow that fell like last century's angels—
> And the starving horses, their shanks brittled over with ice—
> And the moon atop its brilliant derrick, and the poor burning so beautifully in the oilfields.
> As we drove, their cries lit the wind with wailing
> and you said, This isn't America into the truck's dark cab and turned the radio loud.
> —"We Wanted to Find America," Kevin Prufer (2008)

Given the 1990s' willful self-exile into the comfortable nullities of cable television's domestic theater, it was perhaps inevitable that the deep trauma of the terrorist attacks on September 11th, 2001 would be carried live. A stunned nation would watch helplessly as a plane, a spare, thin shadow, headed unerringly, inexplicably straight into the World Trade Center South Tower. A speechlessly short hour later, both iconic towers collapsed in a heavy swirl of toxic ash and debris that made the sun—and our decade long strategy of escape—decidedly ironic. That we watched it live, that a catastrophe of that magnitude would be played out in our kitchens and living rooms, violating our most protected safe-havens, swiftly returned American culture, itself on the sunny threshold of a shiny new millennium, to a most anxious environment. That immediacy, the unblinking feed of television footage that would be endlessly re-looped and then preserved in the permanent syndication of YouTube, made escape non-optional. Given the unrepentant fanaticism of those who had guided the planes into public buildings they saw as military targets and the appalling death toll of what was the bloodiest foreign attack on American soil in close to two centuries, America felt a most unyielding anxiety.

Inevitably, the immediate response to the attacks drew on the familiar rhetoric of the Cold War, specifically the coaxing simplification of its black and white Manichean rubric. Reeling in the aftermath of surprise, we needed certainty. We had in a single, stunning morning moment lost our taste for irony. We needed clear heroes, we needed clear villains, we needed emotions such as sorrow, rage, faith, patriotism. Fittingly, we found our heroes both in the vast corps of gallant first responders—fire fighters, EMS squads, police, hospital emergency staffs—and in the casualties, the nearly three thousand dead whose harrowing experiences were replayed through the raw immediacy of their own desperate text messages, email missives, and telephone calls. And villains, they were certainly clear. Within weeks of the attacks, under the orchestrated efforts of the Bush White House, we had conceived of a compelling and frighteningly vast Axis of Evil, cleanly redrawing the complexities of the international geo-political landscape with bold and easy Crayola strokes. We demonized the Middle Eastern Other, casting Osama Bin Laden, the Saudi-born terror tactician who had apparently masterminded the attack, as the ubiquitous, elusive, perpetually threatening Bogeyman (reflected nearly a decade later, after his bloody execution by an elite Navy SEALs operation, in the unseemly clamor for indisputable photographic evidence that he was indeed dead). We further embraced the certainty and clean logic of a military response, a full-throttle campaign that would render, we were certain, the shadow-enemy verifiable, definable, locatable, and supremely beatable—we went first into the mountains of Afghanistan to cripple Al Qaeda and then within a swift few months into the oilfields of Iraq to destroy that rogue nation's weapons of mass destruction (recall the therapeutic satisfactions first of the flight-suited Bush Victorious on the deck of the *USS Abraham Lincoln* posing beneath the banner MISSION ACCOMPLISHED and later the shaky handheld video feed of Saddam Hussein's carefully staged execution-event). At home, we were determined to eliminate surprise, determined to regain the clarity of causality and the reward of careful logic through an ambitious program of homeland preparations, endorsing a vast spider-web of government oversight and bold security measures at airports, border crossings, and government installations, all the while vigorously quashing dissenting opinion as un-American and cowardly, a movement toward certainty climaxing in the emergence late in the decade of the bumper sticker simplifications and feisty beer-commercial rhetoric of the Tea Party.

It wasn't, as many commentators pontificated, that September 11th was somehow beyond the reach of language, beyond the reassuring logic and discipline of sentence and paragraph. On the contrary, we talked our way through the trauma-event, aggressively working language to neutralize the nagging anxiety that the surprise attack was trauma without meaning. That initial response reflected the appropriation of the rhetoric, sensibility, metaphors (and paranoia) of old school Cold War anxiety—as Aaron DeRosa has so cogently argued. We skillfully indulged the clean-edged, unironic division of the world into good and evil; we re-assumed our traditional sense of largeness, our cornerstone perception of our own grandeur and privilege, the feeling that here something cosmic in scope had been irreversibly altered (Europe, for all its compassionate sympathy extended after the attacks, must surely have wondered over such a grand national reaction to what was basically the destruction of two business offices when land and air battles during World War Two had razed entire cities and when for decades their cities had been subjected to the random acts of large-scale terrorism). We reanimated a conception of evil as a vicious but ultimately predictable agency and a concomitant commitment to the logic of judgment and retaliation unmediated by the subtler arguments of tolerance and mercy.

Anxieties grew. Despite the emergence of the vast bureaucracy of homeland security, and our much publicized and very visible campaign to secure our transportation systems and our borders, we pondered what would happen if (when?) that entrenched fanaticism ever secured nuclear capability or some massive computer virus that could conceivably disrupt all private communication signals, compromise military security systems, cripple financial institutions, and wreak havoc within the operations of government at all levels. But it was more than our anxieties over terrorism. We seemed steeped in a wider environment of crisis. There was the increasingly strident outcry over the measureable impact of global warming as well as the unsettling manifestations of meteorological catastrophes from earthquakes to volcanic eruptions, tsunamis to hurricanes. In addition, there was the increasing evidence of volatile European economic markets that threatened, by late decade, to send governments (and financial markets) into toppling domino chaos, that looming catastrophe counterpointed here by the slow-simmer economic meltdown in the housing, automobile, and banking industries, that discontent expressed in the Occupy Wall Street movement. Then, with mounting concern, we watched as long-entrenched dictatorships in political hotspots along the Middle East rim gave way to the uncertain energy of street visionaries chanting for radical change. At home, despite the celebratory optimism that greeted the incoming Obama administration, by its midterm the country was riven by bickering political bipartisanship and rhetorical rancor that pitted Left against Right, an uncivil war that amped up the uneasy feeling that government itself had become ineffectual and useless. It surely came as no surprise when on January 10th, 2012, no less than the august *Bulletin of Atomic Scientists* announced that it would, for the first time in years, move its Doomsday Clock forward, to just five minutes before midnight, that fictitious moment that since the deepest chill of the Cold War had represented the point at which humanity would self-immolate.

Historically, it is just such a tense cultural environment that produces texts animated by the compassionate vision of Nuclear Criticism. Indeed, an accidental conspiracy of writers in the decade after the attacks offered works that positioned the anxieties of the new millennium within the passionate narrative imagination that since the Book of Daniel had ministered to a people caught within the press of grand crisis. Much like the texts produced in the first generation after Hiroshima, the shimmering novels of Vonnegut, DeLillo, Pynchon, Malamud, and Percy, these dense texts claim a far more provocative and necessary privilege. They suggest the villains in such an anxious age are not some mountain-dwelling, quasi-military terrorist cells but rather facticity itself, the hard edge of certainty, the presumption of trust, the assumption in the validity of data that makes inevitable the embrace of activism, the spread of propaganda, and the vociferous endorsement of the hard jingoism of free-rein patriotism. None of these actually help to ease anxieties or brings peace of mind. Much like the texts produced in the first generation after Hiroshima, these dense texts of Nuclear Criticism suggest that we had lost our taste for nuance. These texts articulate a vision of a community determined to confront its own peril, to reclaim uncertainty and anxiety. The genre's rich, paradoxical vision offers, before the cut to black, the stunning vitality of an earned hope. It is the difficult embrace of the joy of being caught in the middest, the urgent consolation of reclaiming both awareness and uneasiness. Like John of Patmos, these writers counsel embracing the world as a doomed thing and challenge us, nevertheless, to live within that anxiety, on the cusp of surprise, to dare to step forward when such a step is freighted with threat. They offer a reminder that only by engaging the moment rife with such anxiety might you split the fearful now and tap into genuine hope.

Post-September 11th texts that fall within the argument of Nuclear Criticism, not surprisingly, test a variety of genres. Much critical attention has rightly been paid to the narratives (and films) that directly treated the events of September 11th (see for instance the pioneering work done already by Richard J. Gray, Martin Randall, and Ann Kenniston and Jeanne Follansbee Quinn, among others). These texts reclaimed through the recuperative power of the imagination strategies for shaping our collective recollection of the trauma of that day even as the day itself faded into historic record. Collectively these works will stand as a national testimony to how language can at such tectonic moments create meaningful response. They are heartfelt, emotionally vivid, and culturally necessary—but other texts in the decade not tied specifically to these therapeutic works also evidence the wide range and grand reach of Nuclear Criticism.

To list a few is but to suggest the reach of that response. It is in the soaring, defiant rhetoric of forward-looking change in Barack Obama's acceptance speech at Denver's Invesco Field in front of more than 80,000 ardent supporters at the close of the Democratic National Convention in August, 2008; it is in the hard ethos of survival at the core of the decade's signature television format, the reality show, where faux-celebrity contestants persevere against demanding circumstances that defy containment, control, or predictability, who must maneuver in the crazy improvisation of a scriptless mediascape; it is in the disturbing peril of the fabled comic book metropolis Gotham City as envisioned by director Christopher Nolan in the decade's top-grossing dramatic film, *The Dark Knight* (2008), a city brought to the edge of calamity, its heroes compromised by moral ambiguity, compelled to navigate against the bold, calculated terroristic strikes of The Joker, unforgettably embodied by Heath Ledger, anarchy without the convenience (and logic) of an agenda, without the explanation of context or the expectation of accountability, the brutal gorgeous rage of pure chaos; it is in welcoming aura radiating from the decade's two most enduring advertising icons—the relentless upbeat can-do chirp of Flo the Progressive Agent and the cool street Zen of the Geico Gecko—both pitching the reassurance of accessible insurance, a tonic reminder of dire circumstances and the need we all face every day to negotiate such unpredictability; it is in the sweeping end-of-empire poems gathered in Kevin Prufer's widely acclaimed 2008 collection *National Anthem* with its darkling vision of a ruined landscape, of abandoned shopping malls and weed-infested highways, and its Whitmanesque affirmation of the position (indeed responsibility) of the artist to offer the consolation of hope (Prufer compares the poet to a kidnap victim, a boy locked in the trunk of a car abandoned in the woods, still talking, the voice resilient and determined amid dire circumstances); it is in the gospel-tinged, multi-Grammy-winning title song of *The Rising* (2002), Bruce Springsteen's twelve song cycle about the attacks, a song that—unlike the rousing, chest-thumping arena anthems inspired by the attacks—is a far more disquieting vision: the story of a city firefighter, ascending the World Trade Center staircase, caught in smoke and confusion but determined to make the difficult next step. The title is a reference as much to the firefighter's ascent up the steps of the crippled building as to the nation's need to step uneasily into the new normal of the new age.

And this does not begin to exhaust the decade's responses that both magnify and amplify the argument of Nuclear Criticism—consider *Up*, Pixar's poignant fable of old age and mortality; the video game world of Halo; the ongoing interactive cultural narratives that are YouTube, Twitter, and Facebook; the visceral impact of Kathryn Bigelow's war film *The Hurt Locker* (2008); the resurrection tracks from Kanye West's

College Dropout (2004); the unlikely late-decade canonization of a mechanically-challenged second string quarterback named Tim Tebow—each of these cultural texts dares us to live without the convenience of certainty, to privilege both awareness and uncertainty, and to accept the heavy burden of catastrophe as the only condition in which hope and community can rightfully, meaningfully prosper. But few texts better exemplify the promise (and viability) of Nuclear Criticism in the post September 11th culture than Cormac McCarthy's soaring mid-decade parable, *The Road* (2006).

IV

> He stood listening. The boy didnt stir. He sat beside him and stroked his pale and tangled hair. Golden chalice, good to house a god. Please dont tell me how the story ends.
> —Cormac McCarthy, *The Road*

It is certainly not to diminish the considerable achievement of *The Road* to suggest that with its recreation of a grim post-apocalyptic West Virginia, its rich visual texturing, and its cloaking claustrophobic feel it recalls the best of the first-season episodes of Rod Serling's *The Twilight Zone*, the pioneering television series whose subsequent careless appropriation as pop culture kitsch belies it as an expression of Nuclear Criticism. McCarthy's novel could serve as a template for the post-September 11th florescence in texts of Nuclear Criticism, even though McCarthy's narrative has nothing to do with the actual events of September 11th. Although McCarthy had long enjoyed major enthusiasm from academics, *The Road* was the reclusive author's first bestseller with more than a million copies sold in its first months. More remarkable was its enthusiastic imprimatur by Oprah Winfrey who selected this provocative Beckett-esque allegory, with its nightmare vision of roving gangs of half-naked cannibals and grisly episodes of brutal vigilante violence, for her normally genteel book club. Such widespread interest in McCarthy's novel—its selection for the Pulitzer Prize, for instance, and its subsequent development into an Oscar-nominated film—suggests that despite publishers' market pitch that hyped the work's *Night of the Living Dead* dimension, McCarthy is interested in offering something far more elevating.

Given McCarthy's signature fondness for elaborate Faulknerian prose, excessive gothic violence, and dense nonlinear formal constructions, *The Road* is notably accessible. The plot is tightly executed; the symbols, broad and immediately suggestive; the characters, nameless allegorical cutouts with little back story or complicating psychological nuance; the storyline, fragmented into episodes, most a paragraph or two—in short, as with all texts of nuclear anxiety, *The Road* is best approached as a public-event text designed not to be studied nor to settle into the tomb dust of libraries but rather to be read, to reach a wide and needful audience. Clearly, the themes of Nuclear Criticism are central to McCarthy's vision: the hammer-stroke intrusion of a culture-wide trauma followed by the hissing away into inevitable nothingness; the oppressive reality of our vulnerability; the excruciating pressure of dwindling time; the compelling code of right and wrong; the layered symbolic possibilities of seemingly ordinary objects; the shattering intrusion of the lurid and the grotesque; the smothering sense of paranoia; the determination to maintain the integrity of purpose against the certainty of mortality and the abundant evidence of pointlessness; and the vaunting cosmic sensibility that perceives the limited plane of material experience as radically insufficient to the reach of human aspiration.

As with all visionaries whose texts fall within the genre of Nuclear Criticism, McCarthy is not interested in exploiting the big bang of catastrophe itself (the cataclysm, never exactly identified or given any geopolitical context, happened more than ten years earlier; we are told only that "The clocks stopped at 1:17. A long shear of light and then a series of low concussions").[1] The premise of McCarthy's narrative immediately recalls scores of unsettling (and, yes, at times cheesy) sci-fi storylines: heroic survivors—in this case, a father and son—wandering a not-too-distant, post-apocalyptic Mad Max sort of lawless doomscape, a gray scab-land of ashes and charred ruins that clearly recalls our own world. A father, dying of respiratory failure from years of breathing the nuclear winter ash, his body wracked by a bloody cough, seeks only to protect his ten year old son. Together they struggle day-to-day against the elements, a dreary snowy rain and a fierce cold, to scavenge meager sustenance. The father pushes all their belongings—binoculars, tools, blankets, some tins of food—in a grocery cart with a bad wheel. Together the father and son, half-starved, move along the road, stopping only to poke through cupboards of abandoned houses and the shelves of supermarkets for any cans of food, all the while trying not to fall into the clutches of roving gun-toting marauders and far more terrifying bloodthirsty cannibals intent on harvesting stragglers they catch along the wasteland highways for later piecemeal consumption.

Although McCarthy dutifully provides the Boschian touches expected in post-apocalyptic fictions—a stone wall decorated with rotting heads, "dried and caved with their taut grins and shrunken eyes," posted as a warning against trespassing.[2] There is a remote farmhouse wherein naked unfortunates are chained awaiting their eventual consumption as well as the charred, spitted remains of an infant (his is a narrative with a far different agenda). The quiet love between the father and son centers the narrative, a code of caring that heroically defies the larger concessions to brutality and selfishness. Despite the obvious (and appalling) evidence of cataclysm, McCarthy warns, the threat is to surrender to the certainty of despair—in unguarded moments of recollection, the man recalls how his wife had been unable to sustain the logic of continuity and had dispatched herself years earlier apparently by slicing her wrists with a "flake of obsidian…sharper than steel."[3] The man cautions the son never to envy his mother, her only hope in the end "eternal nothingness."[4] The man now carries a pistol with a single bullet and counsels his son on the cleanest way to commit suicide but only in the advent of most dire circumstances, only if the father has died or has been taken, and then only if the boy himself faces imminent capture or the agonizingly slow death of starvation. The narrative resists the centripetal pull of despair—the father indulges the tender trap of memory only in sleep, his dreams haunted by too-vivid images of a world beyond recovery. The slender plot pivots upon the man's determination that given a bleak autumnal world edging into the forbidding absolute of winter, the two need to walk, one grim mile at a time, the nearly two hundred miles south to the warmth and promise of the Atlantic Ocean.

We follow that mission—as heroic as it is pointless. Along the way the two meet appropriately allegorical figures—The Man Who Had Been Struck by Lightning, The Crazy Man Angry with God, The Little Boy without a Family—each bringing to the narrative broad and easy lessons appropriate to a parable, lessons about how to live—or not to live—in the uncertainty of a doomed world. Through it all, the father reassures the boy that the two of them are special, different, that they are the good guys, that they harbor inside them an inextinguishable fire, an allegorical suggestion of both their moral worth and the stability and value of their love, the need they have each for the other. Along the way, the man and his son share in small, fragile, unexpected moments of exhilarating sufficiency typical of Nuclear Criticism: they stumble on a can of Coke in a vandalized supermarket vending machine and relish the delightful warm fizz of its syrupy sweetness; they chance upon a colony of morel mushrooms, wrinkled and dried, amid the mulch and ash along a river and that night fry them up to spice up their beans; they savor the dry bites of shriveled apples they find in the tangled growth of what was once an orchard; and when they luck on an abandoned backyard underground fallout shelter still stocked with canned vegetables and fruits, after a sumptuous meal, father and son relish the simple pleasures of a haircut and bath.

But no place is safe—the father's justified paranoia keeps the two moving steadily toward the ocean. When after days they finally reach it, like the rest of the forbidding moonscape-world, it promises little, "vast and cold and shifting heavily like a slowly heaving vat of slag and then the gray squall line of ash."[5] The long trip has exhausted the father—even as he works diligently to scavenge a boat left aground off shore, he feels the cold approach of death. He cautions the boy that he must get ready to be alone. When the father dies a few days later, the boy, now confused and terrified, spends three long days and nights standing watch over the body. A stranger appears, the father of an itinerant family, who offers to take the boy in—the man reassures the boy that, yes, they are among the good guys (he assures him that they do not eat children). In tears, even as he promises his dead father never to forget him, the boy makes the commitment to the itinerant family, a slender community defined not by blood but rather jerry-rigged from a reckless expression of intuitive trust. The mother greets the boy with a welcoming embrace and a prayer. We are poised for resolution. Then…

Cut to black.

As readers, we are left exposed, vulnerable within the unsettling generous extra-textual space of absolute mystery, uncertain how to feel. Is the boy safe? Is the family what it appears to be? Where will they go? As with so many texts explored in Nuclear Criticism, we are left suspended uneasily between and amid resolutions, dropped rudely into the wide animated energy field of a difficult and dense "perhaps," pitched into the formidable ironic matrix of "possibility." And that must be sufficient. McCarthy argues that all any of us has is the heroic faith to trust against the odds, to go on into the uncertain, confident only that, ourselves doomed, we move about a doomed world. The voiceover authority, the parable-teller, reminds us of as much in the quick-cut to the novel's radiant closing, a paragraph prose poem that recalls brook trout breaking the amber surface of mountain streams. "In the deep glens where [the trout] lived all things were older than man and they hummed of mystery."[6] It is a vision as old as Ecclesiastes that reminds us, despite our anxieties, nuclear or otherwise, the earth—its mysteries, its breadth, its sheer animation—will abide forever.

V

Lorne Michaels (turning to Rudy Guiliani): "Can we be funny?"
—Season Premiere, Season 27, *Saturday Night Live*, 29 September 2001

Ironically, even as I am typing this, lines that assume some sort of posterity, real, if unnamable readers, the world is once again poised to end; this apocalypse is scheduled to occur in just a few months according to the Mayan calendar, a hoary calculation that involves both complex astronomical data and intricate numerology. Of course, given that we all just survived an apocalypse last spring—one predicted by Reverend Harold Camping of Family Radio, a revelation that, as it turned out, required some recalibration—it is easy to be glib, cynical, smug, dismissive, to echo Jeff Tweedy, the quirky singer-songwriter who fronts the indie band Wilco, "Come on, children, You're acting like children / Every generation thinks it's the end of the world."[7]

As it turns out, that's exactly what we need to think. Nuclear anxiety—and Nuclear Criticism itself—has never much been distracted by the elaborate calculations of predicting The End. Mathematical exercise in faux-certainty misses the argument of these most compassionate works. Anticipation is the given—it is the quality of that anticipation, the dynamic of that expectation, that provides the genre its lucid, luminous logic. We cannot not afford to think we are at the end of the world. We cannot not afford to be vigilant. At one point in *The Road*, the young boy awakens whimpering from uneasy sleep and seeks the comfort of his father's arms, unsettled by a particularly vivid dream about a wind-up penguin, a toy designed to delight by mimicking the easy magic of animation, to flap its flippers and even waddle. But in the child's dream, no one had wound the toy—neglected, forgotten, it had terrified the boy and had jolted him from sleep. Like all the parables McCarthy imbeds in his narrative, the implications of the boy's dream are transparent. Yes, McCarthy argues, wonder, the unsuspected magic, the consolation of simple animation, is available, near, within reach—but it requires attention, care, constant vigilance, loving diligence. It is the difficult affirmation of the texts that make up the canon of Nuclear Criticism, a vast body of wisdom literature that reaches beyond any single traumatic historic event. It extends that complicated hope to a humanity that is in perpetual crisis, terrified of the very sense of mystery that alone provides a doomed world its dimension, its nuance, its genuine shock and awe. As David Chase counseled his anxious post-September 11th culture: until the cut to black, don't stop believin'.

Notes

[1] Cormac McCarthy, *The Road* (New York: Knopf, 2006), 52.
[2] Ibid., 91.
[3] Ibid., 58.
[4] Ibid., 57.
[5] Ibid., 215.
[6] Ibid., 287.
[7] Wilco, "You Never Know," *Wilco*, Nonesuch, 2009, compact disc.

CHAPTER SIX

THE PIXILATED APOCALYPSE: VIDEO GAMES AND NUCLEAR FEARS, 1980-2012

WILLIAM KNOBLAUCH[1]

Introduction

In 1983, Americans flocked to see the major motion picture *Wargames*. The film featured Ally Sheedy and Matthew Broderick as teenagers who, attempting to download a video game entitled "Global Thermonuclear War," accidentally hack into a government computer and bring America to the brink of nuclear apocalypse. Like its most famous nuclear war-themed film precursor *Dr. Strangelove: Or How I Learned to Stop Worrying and Love the Bomb* (1964), *Wargames* suggested that human error might trigger a nuclear exchange. Like another Cold War classic, *Fail-Safe* (1964), *Wargames* hinted that technical malfunctions— this time through a computer named "Joshua"—could also cause nuclear Armageddon. While it played on old tropes, *Wargames* became an iconic 1980s film due to its memorable climactic scene. Set in the North American Aerospace Defense Command (NORAD), the film ends as Sheedy and Broderick attempt to convince military generals that the incoming nukes are not real; the computer, "Joshua," is only playing a game. NORAD's screens flicker with atomic explosions as Joshua cycles through every possible attack scenario, only to conclude that nuclear war cannot be won. The film ends with Joshua robotically commenting to its creator, Professor Falken, on the futility of nuclear war: "A strange game. The only winning move is not to play."[2]

The Pixilated Apocalypse: Video Games and Nuclear Fears, 1980-2012   123

Fig. 6-1: Screen capture from the 1983 film *Wargames*, set in NORAD.³

Wargames' merging of nuclear fears and video game culture was timely. By the late 1970s, video games had become a ubiquitous part of American youth culture. Arcades littered downtowns and strip malls; coin-operated machines appeared in American pizzerias, gas stations, and taverns; the Personal Computer (PC) and home video game console markets exploded, bringing games into millions of homes. As graphics improved, game characters—such as Donkey Kong, Mario, and Pac-Man—became iconic, but most digital games were militaristic in nature. This video game revolution coincided with an escalation of nuclear fear and American militarism. The 1979 Soviet invasion of Afghanistan, the nuclear reactor meltdown at Three Mile Island, and the 1980 election of Ronald Reagan—with its resulting nuclear arms buildup and American militarism abroad—sparked new nuclear concerns. In response, antinuclear demonstrations, such as the Nuclear Freeze campaign, reached record numbers. In the early 1980s, the concurrent rise of video game culture and nuclear fear made *Wargames* both popular and pertinent.

Wargames was only part of a 1980s pop culture that reflected America's growing uneasiness with nuclear weapons. In 1982 nonfiction books like Jonathan Schell's *Fate of the Earth*, Ground Zero's *Nuclear War: What's In It For You?*, and *Freeze!: What You Can Do to Stop the Arms Race* graced *New York Times* bestseller lists. Antinuclear pop songs like Nena's "99 Luftballoons" (1983), Frankie Goes to Hollywood's "Two Tribes" (1984), and Sting's "Russians" (1985) all warned of the apocalypse on FM dials. The films *Testament* (1983) and *Threads* (1984) depicted postwar horrors on the big screen, and the antinuclear television docudrama *The Day After* (1983), about a nuclear attack in Lawrence, Kansas, achieved record ratings. Popular scientist Carl Sagan took to television and warned of the "nuclear winter," a post-nuclear war scenario in which smoke would block sunlight, kill vegetation, and end humanity. These are but a few examples of a robust 1980s antinuclear popular culture, and video games were not immune to this atomic trend.[4]

This essay examines the connections between video games and cultural perceptions of the nuclear threat since 1980. Created during the Cold War, early video games adopted militaristic themes including space warfare, tank battles, and hand-to-hand combat. The most popular games were "shoot-em-ups," games in which players tried to shoot and destroy as many enemies as possible.[5] By the 1980s, video games began appropriating nuclear war for gameplay, and game designers faced the problem made explicit in *Wargames*: What fun is a game in which the only way to win was not to play? The answer to this question changed over time. During the 1980s, a decade of heightened nuclear fears, video games utilized nuclear war fighting scenarios as plotlines, adopted current ideas about nuclear defense for gameplay, and lambasted Cold War political leaders—such as Ronald Reagan—who might trigger the apocalypse. In the post-Cold War 1990s, video games mostly abandoned these themes. While 1980s games featured themes of fighting or preventing a nuclear war, 1990s games used the imagined post-apocalyptic landscape as a narrative backdrop. After 9/11, fears of terrorist nuclear attacks combined with a video game industry trend towards first-person shooters, resulting in new, immersive games that incorporated tactical nuclear weapons as useable gameplay elements. While nuclear themes in video games have changed over time, they have consistently reflected broader cultural assumption about the global nuclear threat.

Scholars and cultural critics have added much to our understanding of atomic popular culture, yet video games remain an under-studied media.[6] This oversight is all the more glaring considering video games' commercial prominence. In 2011, the video game industry netted around $20 billion, an amount that surpassed the combined profits of the motion picture and music industries *combined*.[7] The scale and scope of video games is immense, and with so many games produced since 1979, any analysis of the genre needs limits. This essay focuses on games in which nuclear strategy, weaponry, or culture, are central to gameplay. Games such as *Nuclear Strike* (1997) or *Nuclear Dawn* (2011) sound nuclear-themed, but these are, respectively, helicopter and first-person shooters.[8]

While the *Metal Gear Solid* series features antagonists seeking nuclear weapons, and even warhead storage facilities as playable levels, these elements are only narrative backdrops to otherwise traditional stealth games.[9] Games included in this analysis, such as *Missile Command* (1980), *Theatre Europe* (1985), and *Fallout 3* (2008), all rely on nuclear strategy, weaponry, or science (i.e., radioactive fallout) as core gameplay elements. Since 1980 nuclear war has proved to be a surprisingly durable theme for video games. As these examples will show, video games have represented larger cultural appraisals of the nuclear threat and atomic anxieties of the decades in which they appeared.[10]

The 1980s, Nuclear War and Nuclear Defense

Video games were a product of Cold War science and technology. Manhattan Project member William Higinbotham arguably designed the first video game prototype, *Tennis for Two* (1958). Early video game titles, like *Spacewar* (1962) and *Combat* (1977), set the stage for a medium dominated by violence and warfare. Even video game pioneer and *Pong* creator Ralph Baer's 1970s game prototypes had military applications, such as his "Interactive Video System Rifle Training System" and "Light Antitank Weapon Simulator."[11] Yet video games did not begin to depict nuclear war until the resurgence of nuclear fears in the early 1980s. After a nadir of antinuclear activism in the 1970s, nuclear tensions escalated dramatically during Reagan's first term.[12] In addition to the Soviet war in Afghanistan and Reagan's own nuclear arms buildup, geopolitical events of 1983 added to global tensions. That year, a deadly terrorist attack on U.S. Marines in Lebanon, the U.S. invasion of Grenada, the Soviet shoot down of Korean Airliner 007, the deployment of U.S. nuclear warheads into Europe to support NATO nuclear forces, and continued U.S. covert operations in Soviet-occupied Afghanistan, made Reagan's first term the tensest Cold War period since 1962s Cuban Missile Crisis. In the early 1980s nuclear fears were real, and many of the era's video games reflect the belief that nuclear war was indeed possible.[13]

One of the most popular coin-operated (coin-op) games of the early arcade era was Dave Theurer's *Missile Command* (1980). Created in the shadow of increased superpower animosity, Theurer envisioned a realistic game in which players protected cities along the California coastline. Gameplay would be simple and addictive. Players utilized a track-ball, not the traditional joystick, to target and intercept incoming Intercontinental Ballistic Missiles (ICBMs) from a ground-based Anti-Ballistic Missile system (or ABM). As levels progressed, these nukes descended in greater speed and numbers. Inevitably, players lose all their cities, but instead of a "Game Over" screen, *Missile Command* features an ominous explosion and a flashing "THE END" message.[14] According to Theurer, the game concludes with "THE END" so that players would realize a "final lesson... [that] nobody wins in a nuclear war." Theurer remembers that he "wanted people to become aware of the horrors of a nuclear war. I didn't want to put players in the position of nuking entire cities as entertainment, because it would desensitize them from such horrors."[15] *Missile Command's* popularity suggests that defending against a nuclear war might be more appealing than waging one, an opinion President Reagan shared. In March of 1983, Reagan announced a plan to create a new antinuclear ballistic missile system. Most American scientists scoffed at this futuristic idea, which might involve "lasers in space." Regardless of its plausibility, Reagan's Strategic Defense Initiative (SDI, or "Star Wars" to skeptics) captured the public's imagination. By 1985, the films *Spies Like Us* and *Real Genius* (both 1985) appropriated SDI programs as plotlines; it took even less time for video games to adopt Reagan's program.[16]

One early SDI-based game was *Wargames*. Based on the 1983 film, *Wargames* opens with a "Greetings Professor Falken" message from the "Joshua" computer which informs gamers that they are about to defend against a "Global Thermonuclear War." In *Wargames* players control an SDI-like system to intercept incoming missiles, airplanes, and bombs. Players can choose between fighter planes, "ABMs" to intercept incoming ICBMs, and even a "satellite weapon" which can fire laser beams. As attacks intensify, the game's DEFCON meter increases. At its highest alert level, DEFCON triggers a global thermonuclear war and the end of the game. *Wargames* plays like *Missile Command*, but it bypasses the movie's overriding message that a waged nuclear war cannot be won. Instead, the game adopts a pro-SDI message in which America can successfully defend against a nuclear attack.[17] Other SDI-themed games followed. By 1987, the turn-based strategy game *High Frontier* simulated "the development and actual use of the SDI system" and allowed players to fund research, develop technology, and ultimately deploy an SDI system. In 1987, Sega released *Strategic Defense Initiative*, a coin-op in which players navigate a satellite through an "offensive" stage to fire lasers and intercept ICBMs, and a "defensive" stage similar to *Missile Command*. SDI's popularity began to wane after 1987, but even if SDI hype was short-lived, Reagan's idea proved quite durable in the realm of video games.[18]

SDI was part of Reagan's broader program to reassert American confidence during the Cold War. In his speeches, Reagan repeatedly asserted America's moral superiority over the Soviet "Evil Empire" while U.S. forces battled in Lebanon and Grenada, and covertly supported anticommunists in Latin America and Afghanistan. This combination of rhetoric, military adventurism abroad, and Reagan's arms buildup influenced 1980s popular culture. Films such as *Red Dawn* (1984), in which high school students combat a coordinated Soviet-Cuban invasion of America, was an aggressive (if juvenile) cultural affirmation of American military might and moral clarity. G.I. Joe figures, with their fatigues and high-tech weapons, returned to toy store shelves and, in 1982, became a popular syndicated cartoon. That same year Sylvester Stallone played John Rambo, a troubled Vietnam vet harassed by an ungrateful American police force. By the 1985 sequel to *First Blood*, the awkwardly titled *Rambo: First Blood, Part II*, Rambo is on the offensive. He returns to Vietnam, rescues Prisoners of War (POWs), and single-handedly decimates entire battalions. Rambo's most memorable line, "Can we win this time?," reflects Reagan's desire to shake off the 1970s "Vietnam Syndrome" and restore American military optimism. These examples show how this Cold War offensive permeated 1980s pop culture.[19]

In the Reagan Era, Americans wanted to see its heroes win wars, even if they were nuclear wars, and many 1980s nuclear-themed games were offensive in nature. *B-1 Bomber* (1980), *B-1 Nuclear Bomber* (1982) and *Raid Over Moscow* (1984) simulated tactical assaults on the Soviet Union. *B-1 Bomber* was a text-only game: it contained no graphics; players simply typed commands for gameplay. Still, *B-1 Bomber* presented a plausible Cold War scenario of nuclear escalation and retaliation. Players "flying a B-1 bomber…over the arctic" must locate their primary bombing targets and attack. A sequel, *B-1 Nuclear Bomber*, kept the atomic attack premise but added graphics. *Raid Over Moscow* (1984) challenged gamers to guide their aircraft, intercept incoming Soviet bombers, land safely within the U.S.S.R., and finally invade a Soviet nuclear facility. *Theatre Europe* (1985) simulated a Warsaw Pact invasion of the North Atlantic Treaty Organization (NATO) countries of Western Europe, as did Mindscapes' *Balance of Power* (1985) which promoted gameplay based on "Geopolitics in the Nuclear Age." With actual American nuclear warheads supporting real NATO forces in Western Europe, these nuclear war-fighting video games resembled real-life geopolitical possibilities.[20]

Some 1980s nuclear games were less serious, opting instead to lampoon Cold War leaders. *Nuclear War* (1987) featured thinly-veiled caricatures of American President "Ronnie Raygun," Cuban dictator "Infidel Castro," Libyan Colonel "Khadaffy," and Soviet leader "Gorbachef." Even past U.S. presidents "Trick Dick" (Nixon) and "Jimmy Farmer"

(Carter) make appearances. *Nuclear War* is a turn-based strategy game in which players build up their nuclear arsenals, gain popular support, and launch ICBMs to kill opponents. Despite its nuclear theme, *Nuclear War's* comically drawn characters and schoolyard dialogue keep things lighthearted. Ayatollah "Kookamaine" (Khomeini), for example, might respond to your thermonuclear attack by boasting that "you couldn't hurt a fly." *Nuclear War* even pays homage to Stanley Kubrick's classic atomic film, *Dr. Strangelove.* The game's title scene re-creates Kubrick's Lieutenant Kong riding an atomic bomb rodeo-style as it descends to its target.

Fig. 6-2: Screen capture from *Nuclear War*, which pokes fun at real world political leaders.

The 1988 game *Spitting Image* also features unflattering representations of political figures. Adopting imagery from the popular British TV series of the same name, *Spitting Image* opens with a tongue-in-cheek apocalyptic scenario: "Within the next seven years, a world war greater than all other wars will take place. A war so great even the Swiss will get involved this time. A war so terrible that the Italians have already surrendered." The introduction fades to a globe spinning in space, the earth shaped like a lit bomb. Instead of a turn-based global strategy game, *Spitting Image* is a "beat-em-up." Players can control Ronald Reagan, Margaret Thatcher, Mikhail Gorbachev, Ayatollah Khomeini, or Pope John Paul II. Each of these characters is lampooned in trademark *Spitting Image* fashion, with grossly distorted features: Gorbachev's birthmark animatedly spins on his forehead while Reagan wears a clown suit and big red nose. *Spitting Image's* final battle features a mystery combatant who, once defeated, is revealed to be that fictional 1980s icon of hyper-American militarism, John Rambo.

Whether they were serious or comical, text-based or graphical, strategy themed or combat fighters, these 1980s video games appropriated nuclear war fighting scenarios for gameplay. During a decade in which ideas about space-based missile defense captured the public's imagination, games incorporated SDI or "lasers-in-space" elements. Other games criticized Cold War leaders, presenting them as bumbling commandos awkwardly waging a nuclear war or brawling hand-to-hand. Overall, 1980s atomic-themed video games presented nuclear war as a contemporary issue. Nuclear war as depicted in these games did not take place in the 1950s or during the Cuban Missile Crisis; instead of a threat from the past, these games presented nuclear war in the here and now. In the 1980s, the pixilated apocalypse was a current affair. Such contemporary depictions of nuclear devastation would quickly disappear with the presumed end of the Cold War in the early 1990s.

The 1990s, Atomic Apocalypse and Nuclear Nostalgia

The fall of the Berlin Wall in 1989, the end of communist rule in Eastern Europe, and the Soviet Union's unexpected disillusion in 1991 marked the theoretical end of the Cold War. Shifts in 1990s popular culture reveal how this tectonic shift in global geopolitics signaled the presumed end of American atomic anxiety.[21] During the Cold War, atomic culture featured radioactive mutants including superheroes (*X-Men*), giant lizards (*Godzilla*), and even oversized ants (*Them*).[22] After the Cold War, American culture kept sci-fi themes of mutation but dropped their radioactive aspects. Comic book icon Spiderman is a prime example: Peter Parker originally gained his powers from a radioactive spider bite; in the 1990s film reboot, that fateful spider is *genetically* mutated. The apocalypse remained popular in the 1990s, but films such as *Deep Impact* and *Armageddon* (both 1998) abandoned nuclear war and adopted incoming meteors as humanity's ultimate threat. Video games mimicked this trend of post-Cold War nuclear amnesia. In 1998 two games based upon *Wargames* appeared (both entitled *Wargames: DEFCON 1*) but instead of a Soviet-launched nuclear scenario, players battle against a defense department computer seeking global domination. When atomic themes appeared in pop culture, they came across as comical, lighthearted, and unthreatening aspects of a bygone era.[23]

The "nuclear winter" hypothesis provides one example of how 1990s video games playfully repackaged 1980s atomic science and culture. First introduced in 1983 by popular scientist Carl Sagan, "nuclear winter" is a scientific hypothesis suggesting that fires from a nuclear war would produce masses of airborne particulates that would rise into the atmosphere, encompass the globe, block sunlight, and trigger human extinction. The term "nuclear winter" proved to be a durable addition to the atomic lexicon. By joining two previously unrelated terms, "nuclear winter" conjured up new imagery that combined atomic power with notions of coldness and grayness. The term was colorful, provocative, and durable. In the 1990s, however, "nuclear winter," like many aspects of Cold War atomic culture, lost its fearful resonance.[24] The term re-emerged in numerous 1990s video games, for example in the atomic-themed "shoot-em-up" series *Duke Nukem* (1991). Featuring a crass, flat-topped protagonist who battles against "Dr. Proton" and his horde of radioactively-mutated henchmen, *Duke Nukem* received an upgrade in *Duke: Nuclear Winter*. This 1997 expansion pack features antler-adorned baddies, demonic snowmen, and gun-toting Santa Clauses. Obviously, Sagan's scientific theory did not influence *Duke Nukem's* gameplay, but the term "nuclear winter" conveys the game's wintry but atomic setting. Other 1990s games, such as the first-person shooter *Half-Life* (1998) and the racing game *Treadmarks* (1999), offer similar "nuclear winter" expansion packs featuring wintery terrain. The game *Freedom Force* (2002) appropriates the term for a villain named "Nuclear Winter," a Soviet arms smuggler who dons winter apparel and has the ability to freeze opponents. In each example, the term "nuclear winter" lost its Cold War connotations. Instead the term conveyed some combination of explosive power and a wintery setting. Such was the greater trend of post-Cold War atomic culture: terms like "nuclear winter" never wholly vanished, but they lost their fearful connotations.[25]

1990s video games also continued to use the "last man standing" trope popular during the Cold War. In the 1950s, films such as *Five* (1951) and *The World, the Flesh, and the Devil* (1959) brutally depict survivors' struggles after the bomb is dropped. Into the 1970s, post-apocalyptic survivalism endured, and low-budget films including *A Boy and his Dog*

(1975) and *Mad Max* (1979) feature "last man standing" heroes.[26] Reflecting American beliefs that individuals could live through the apocalypse, survivalist literature endured throughout the Cold War, and even experienced a modest resurgence in the atomically-tense 1980s. Examples include Books entitled *Life after Doomsday* (1980), *Survive the Coming Nuclear War* (1984), and *Nuclear War Survival Skills* (1987).[27] When it came to video games, the most influential atomic narrative was Walter Miller's sci-fi classic, *A Canticle for Leibowitz* (1959). It told the tale of underground monks who, centuries after a nuclear apocalypse, work to preserve human knowledge as society rebuilds itself. *A Canticle for Leibowitz* helped to establish now classic apocalyptic themes—a Luddite movement rejecting evil science, wandering hordes of wasteland survivors, and above-ground radioactive mutants.

Miller's ideas reemerged in PC games. By 1988, the game *Wasteland* promised "adventure in post-nuclear America." Like *Canticle for Leibowitz*, the game features radioactive mutants, roaming gangs of thugs, and libraries in which players can learn valuable forgotten survivalist skills, such as metallurgy; players can even purchase survivalist items like guns, ammo, and Geiger counters. In short, *Wasteland* combines sci-fi with survivalism and set a standard for post-apocalyptic gaming. Like the most outlandish 1980s survival guides, *Wasteland* suggests that Americans who had the right tools, gadgets, and wherewithal could thrive in the post-apocalypse.[28]

Longstanding notions of American exceptionalism would continue to influence digital fantasies about dominating the post-apocalyptic frontier. In 1993, European game developers at Max Design released the bleak atomic-wasteland survivalist game *Burntime*. The game's characters remain traumatized by the holocaust. Some radioactive mutants stay friendly; others threaten to eat you. In *Burntime*, the goal is to rule the world (or at least what's left of it). To do so, players must survive the elements, salvage wasteland scraps, build rattraps for food, find potable water, and bargain at "Atomic Pubs" in otherwise decimated towns. Illustrative cut-scenes feature a burnt, bleak landscape. *Burntime's* imagery connects the pixilated apocalypse with 1990s heightened concerns over global environmentalism. Game magazine *CU Amiga* called *Burntime* "one of the most environmentally aware games every released." Such awareness can be attributed to environmental group *Greenpeace's* contributions to the game's manual. 129 pages in length, *Burntime's* manual devotes its first seventy-eight pages to detailed explanations of radioactivity, a historical overview of the 1986 Chernobyl disaster, the science of the greenhouse effect, and the threat of global climate change.

A chapter entitled "Water: On the Way to Burntime" informs gamers about wasteful water practices and advises them not to "be one of the sheep... do without a green lawn that requires intensive watering. Plant a field of flowers with a garden pond. Remind yourself that water is not inexhaustible. Treat it with a sense of responsibility!" This combination of environmental activism and apocalyptic gameplay was telling. *Burntime* kept a Cold War sci-fi narrative of the "last man standing," but combined it with 1990s environmentalism.[29]

This combination of atomic culture and global environmentalism emerged in *Wasteland* and became clearer in *Burntime,* but 1997s *Fallout* raised depictions of the post-apocalypse to new levels. More than any other video game franchise, the *Fallout* series combined Cold War culture and science with a dangerous post-nuclear war environment. *Fallout* and its sequel *Fallout 2* (1998) both included detailed instruction manuals rich with atomic iconography. *Fallout's* in-game cut scenes explain the cause of the apocalypse: a global struggle over increasingly scarce natural resources, in particular petroleum and uranium. *Fallout's* manual ("The Vault Dweller's Survival Guide") is rich with 1950s imagery and provides detailed information about "nuclear blast effects" and the "atmosphere effects" of mushroom clouds reminiscent of early Cold War U.S. Civil Defense propaganda. For example, the manual explains that nuclear blasts create "heat from fusion and fission [which] instantaneously raises the surrounding air to 10 million degrees C." Despite these stark warnings, the manual's mascot, "Vault Boy" maintains his unfailing grin, smiling even as he ducks and covers under his desk—a clear nod to early Civil Defense cartoon icon "Bert the Turtle."[30]

Yet even with its reliance upon early Cold War themes and imagery, *Fallout* keeps with the 1990s trend of replacing radioactivity with more modern concerns. Much as *Spiderman* had shed his radioactive back story, *Fallout's* protagonist must ultimately combat mutants who were transformed not by radioactivity, but from being dipped in genetically-mutative materials.[31]

1990s games suggest that after the Cold War, themes about nuclear defense, technology, or strategy, resonated little with gamers. Nostalgic takes on nuclear war remained, but only if they included more contemporary themes of environmental degradation or genetic mutation. When games appropriated atomic science or terminology, such as "nuclear winter," they did so with little regard for these term's original Cold War meanings. This imagery, such as the type used in *Fallout*, remained kitschy and unthreatening—a far cry from 1980s nuclear war simulators and SDI-themed games. In the 2000s, the landscape of nuclear games would change yet again, and many games included something new in their pixilated apocalypse: detailed atomic explosions and nukes as not as signals of the apocalypse, but as viable weapons of war.

Fig. 6-3: "If you see the flash, duck and cover!" "Vault Boy" from *Fallout*.[32]

Video Games and Nuclear Weapons After 9/11

A tragic watershed moment in history, the terrorist attacks of September 11th, 2001 shocked the world and altered the course of American history, politics, and culture. 9/11 signaled a return of American atomic anxiety, albeit of a different kind. Pundits and strategists agreed that a *Wargames*-like scenario of global thermonuclear war was reduced, but the threat of isolated terrorist attacks with nuclear weapons or radioactive "dirty bombs" had increased considerably.[33] Much as they had in previous decades, video games reflected larger cultural ideas about the nuclear threat. In the 2000s, atomic culture emphasized political and military factionalism in the post-apocalyptic world. Cormac McCarthy's novel *The Road* (2006, with a film adaptation, 2009) provides a bleak, "nuclear winter-like" landscape in which roaming hordes of cannibals hunt innocents.[34] The TV series *Jericho* (2006-2008) depicts an American town's struggles after a nuclear attack. As pop culture rehashed Cold War tropes of domestic atomic attack, abroad the U.S. military continued its actions in Iraq and Afghanistan.[35] Increased U.S. military involvement in the Middle East coincided with the dominant trend in video games—the first-person shooter, or point-of-view games that emphasized tactical military strategy. Such games were more detailed in gameplay and visceral in violence. By removing players from the "God's eye" view and inserting them into a first-person apocalypse, the message had changed. Now, instead of defending against a national or global attack, gamers only had to worry about themselves, not the survival of mankind. 1990s atomic-themed games largely lacked this first-person point of view, and games with turn-based mechanics seem clunky when compared to the fast-paced and visceral experience of 2000s first-person shooters.

Although it continues to use 1950s nostalgia, the third installment of the *Fallout* series (2003) reflects trends in the 2000s gaming landscape. Set in the post-apocalyptic wasteland of Washington D.C., *Fallout 3* greatly improves upon its predecessors, combining Cold War kitsch with futuristic technology in startling detail. *Fallout 3*'s manual asks gamers to "imagine if, after World War II... technology progressed at a much more impressive rate, where American society remained locked in the cultural norms of the 1950s." The game succeeds beautifully in presenting this idyllic "world of tomorrow," one filled with servant robots, beehive hairdos, and fusion-powered cars. Dilapidated suburban homes feature 1950s-style transistor radios, but their garages house nuclear-powered cars. *Fallout 3*'s apocalyptic landscape is born of an anachronistic Cold War past in which America and China engaged in a nuclear war. The exclusion of the Soviet Union signals that *Fallout 3* courts a post-Cold War gaming generation, and that old tropes of Soviet communism no longer resonate with gamers. Simultaneously, the game's reliance on 1950s imagery suggests that nuclear war was only ever really possible during the early Cold War. Put simply, *Fallout 3*'s apocalypse is born of a distant, but culturally familiar, 1950s era; it is distinctly different from 1980s video games which simulated nuclear war in the here and now. By the 2000s, nuclear war was only a possibility of the distant, even fictional, past.[36]

Fallout 3 begins in Vault 101, a gigantic bomb shelter that has preserved elements of 1950s culture. Delinquents form a gang, the "Tunnel Snakes," with slicked back hair and leather jackets. Meals are served in a cafeteria resembling a classic American diner, replete with a red-and-white checkerboard floor, vinyl upholstered booths, and a 1950s-era jukebox. Outside the vault, cars flaunt giant tailfins; the Protectron, a wasteland robot, resembles Robby the Robot from the 1956 film *Forbidden Planet*; even the game's soundtrack is decade-appropriate, and features Cole Porter and The Ink Spots. An in-game virtual reality simulation entitled "Tranquility Lane" is a black and white view of 1950s suburbia, complete with a quaint cul-de-sac, picket fences, manicured lawns, and lemonade stands. *Fallout 3's* overarching message is obvious: the nuclear threat is as distant as sock hops and drop tops.

Fallout 3's combination of 1950s nostalgia and realistic atomic hazards accentuate a startlingly detailed, bleak, and scientifically realistic postwar environment. The game's wasteland includes little plant life; the skies are overcast and hazy, and the little life that remains competes for resources. Here, finally, is a sobering representation of "nuclear winter" in a video game. As travelers wander the wasteland, they must monitor radiation poisoning, measured in "rads," the legitimate, scientific measure of radiation intake. Adept players might opt to consume "Nuke Cola" instead of risking health points. Outside the vault, mutants abound, including giant flies, crabs, and cockroaches. In a mini-quest entitled "Those!," *Fallout 3* pays homage to the giant ants of *Them!*, but even the humanoids, or "ghouls," players encounter are freakishly disfigured and mentally anguished in gruesome detail. These are the inhabitants that gamers battle first-hand, and this is *Fallout 3's* biggest shift from its predecessors. The first two *Fallout* games featured turn-based controls. To attack, players simply selected an attack function and waited for the computer enemy to counterattack. Players did not control combat in real time. Like many popular post-9/11 games, *Fallout 3* is a first-person shooter. More than any other genre, militaristic first-person shooters dominated video games in the 2000s, and cultural commentators have connected increased American military actions abroad with visceral violence in home gaming. *Fallout 3's* storyline implies that the nuclear apocalypse is a product of the fictional past, but surviving the wasteland requires modern military tactics. This combination of real tactics in a virtual battlefield has been a cornerstone of post-9/11 first-person shooters. By the 2000s, the U.S. army had even developed first-person shooter games as recruitment tools. In short, *Fallout 3* remains very much a product of post-9/11 video game culture: it emphasizes real time combat and, extreme in its graphics and violence, it strives for battlefield realism.[37]

Fallout 3 embraces another trend in post 9/11 video games—the return of atomic explosions. Unlike the miniature or blurry explosions of

1980s gaming graphic technology (i.e., *Missile Command*), in more modern games these explosions are loud, detailed, and powerful. At one point in *Fallout 3*, players can choose to detonate a dormant nuclear warhead, which has become a religious icon for residents in the wasteland town of Megaton. Detonating the warhead negatively affects player's Karma levels (a trademark of the series), but no doubt the explosion is impressive. While there was one notable atomic explosion in *Fallout 2*, it occurs only after defeating the game's final boss, and players do not have to deal with the consequences of their actions.

Fig. 6-4: Screen capture of detonating a nuclear warhead in *Fallout 3*.

The inclusion of massive mushroom clouds was hardly exclusive to *Fallout 3*. In the post- 9/11 video game landscape, the mushroom cloud returned with a vengeance. Instead of global thermonuclear war scenarios, or nuclear defense technology of the 1980s, post-9/11 games embrace tactical, low-level nuclear explosions as usable and deadly weapons of war. Numerous real- time strategy games, in which players have a "God's eye" view of the playing field, incorporate nukes as highly destructive weapons. The popular *Command & Conquer* (*C & C*) series offers "Nuclear Artillery" and "Atom Rockets" in the sequels *Red Alert 2* (2000), *Generals* (2003), *Tiberium Wars* (2007) and *Red Alert 3* (2008). *World in Conflict* (2007) allows gamers to battle the Soviet Union with "Tactical Nukes." In *R.U.S.E.* (2010), players might choose a "nuclear" mode that incorporates Medium Range Ballistic Missiles (MRBMs). *Supreme Commander* (2007) and its sequel *Supreme Commander 2* (2010) includes tactical nukes, as did Sid Meier's successful *Civilization* sequels, *Civilization: Revolution* (2008) and *Civilization 5* (2010). In each of these real-time strategy games, nukes remain the ultimate trump card—the biggest weapon by which to decimate your enemy.

In addition to these strategy games, first-person shooters of the 2000s began incorporating nukes as integral parts of a game's plotline. *Mercenaries: Playground of Destruction* (2005) tasks players with aborting launched ICBMs. In one level of *Call of Duty 4: Modern Warfare* (2007), players flee from a Middle Eastern city as terrorists detonate a nuclear warhead; the resulting shockwave downs your helicopter. Later in the game, a maniacal madman launches ICBMs towards America, but players reach the missile silo's control panel in time to detonate the missiles mid-flight. In *Modern Warfare 2* (2009), a Submarine Launched Ballistic Missile (SLBM) is detonated in the atmosphere for its Electromagnetic Pulse (EMP) which disrupts military vehicles on the ground. The horror-themed first-person shooter *First Encounter Assault Recon* (or *F.E.A.R.*, 2005) features an atomic blast that sends cars flying through the air. In most cases, players survive these attacks, although not without first witnessing detailed shockwaves and mushroom clouds.

First-person shooters' incorporation of tactical nukes for gameplay coincided with President George W. Bush's promise to diversify America's nuclear arsenal. In 2003, Bush approved a policy directive that suggested using smaller, tactical nuclear weapons in the war on terror, including "mini-nukes" and "bunker busters."[38] Soon, video games included such weapons. The 2008 sequel to *Mercenaries*, entitled *Mercenaries 2: World in Flames* (2008), provides a "nuclear bunker buster" as an important arsenal addition. In *Crysis* (2007), players operate a tactical atomic gun (i.e. the "Tac Gun") as well as an atomic-equipped tank (the "Tac Tank") reminiscent of the real atomic cannon housed at Albuquerque's atomic museum. Intrepid *Fallout 3* players might find a "Fat Man" rocket that fires "mini nukes" at enemies. Both *Modern Warfare 2* (*MW2*) and *Modern Warfare 3* (*MW3*) include death match options that reward players who achieve multiple consecutive kills with tactical nukes. In *MW2*, to use tactical nukes means instant victory for the team. In *MW3*, this "M.O.A.B." (Mother of All Bombs) kills all enemies, while its resulting EMP temporarily disables enemy electronics. In every case, these video game nukes do not wipeout civilizations. They are simply useful weapons of war. The extensive inclusion of nukes in first-person shooters suggests that in the post-9/11 world, the stigma of nuclear weapons has dissipated. Gamers recognize nukes less as a threat to civilization and more as a means of victory.

Conclusion

In the 1980s, gamers played scenarios in which they defended nations against nuclear attack, or battled Cold War leaders in a comic fashion. In the 1990s, players roved the post-apocalyptic landscape, but it was one in which nuclear weapons largely remained relics of the past. By the 2000s, video games have reinforced the Bush administration's opinion that nuclear weapons were again viable weapons of war. As America return to mining plutonium and refining tritium for nuclear warheads, video games present nuclear weapons as welcome additions to military arsenals.[39] These shifts in convention are considerable. Just a decade prior, video games presented nuclear weapons as a relic of the past, a piece of Cold War sci-fi nostalgia. In the 1980s video games presented nuclear war as a contemporary scenario, with nuclear defense technologies—even fanciful ones like SDI—as ideas ripe for appropriation. The more recent video game trend of presenting nukes as useable weapons is both curious and alarming. It suggests that many gamers no longer share the idea made clear in *Wargames:* that when it comes to nuclear war, the only winning move is not to play.

There are signs that video games might someday return to this previously accepted wisdom about the futility of nuclear war. In 2006, the real time strategy game *DEFCON* allowed players to strategize their own global thermonuclear war. The game proudly appropriates the look and feel of *Wargames'* climatic conclusion in NORAD for a whole new generation of gamers; indeed, *DEFCON's* designers admit, their game "came about after watching the movie *Wargames*."[40]

Fittingly, *DEFCON* rehashes the final message of *Wargames*. Players can choose numerous scenarios in which to wage nuclear war, but regardless of the strategy, once the nuclear war starts, nobody will come out unscathed. As *DEFCON's* tagline and webpage boasts, in this game "everybody dies." In a decade in which video games present nuclear weapons as a means of achieving victory, *DEFCON's* bleak message remains rooted in the past, but is arguably much more realistic than its contemporaries. If video games increasingly inform how emerging generations learn and view the world—and contemporary scholarship suggests just that—today's gamers remain largely ignorant to the dangers of nuclear weapons. Nuclear fear, a pervasive reality of the Cold War, now seems to be a relic of the past; unfortunately, the nuclear threat remains very real.[41]

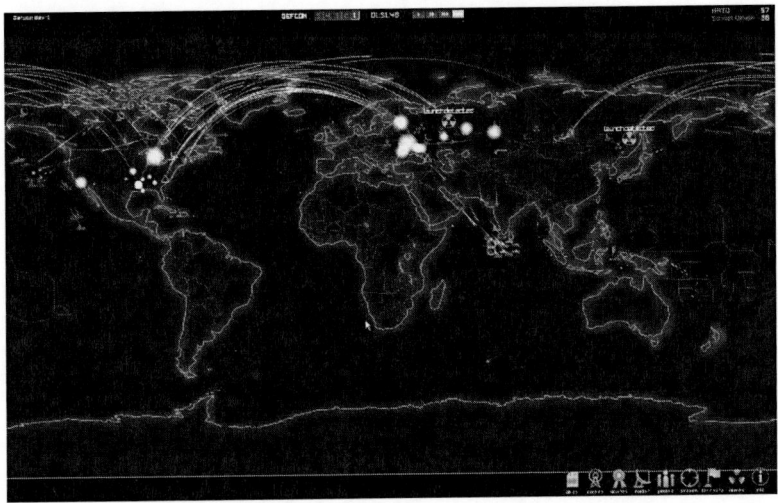

Fig. 6-5: Screen capture from Introversion Software's *DEFCON* (2006).

Notes

[1] John O'Breza contributed much insight regarding the game *Fallout 3*. Mr. O'Breza received his MA in US Intellectual & Cultural History from Ohio University. He is a freelance writer and teacher living in Shanghai, China.

[2] Much thanks to John O'Breza for his contributions on the game *Fallout 3* in this article. David Sirota, *Back to Our Future* (New York: Ballantine Books, 2011), 153; Ed Halter, *From Sun Tzu to Xbox: War and Video Games* (New York: Thunder Mouth's Press, 2006), 94.

[3] Screen capture found at the excellent Cold War cultural repository webpage, Conelrad UK, online at: http://conelrad.com/newswire.php?id=P252, accessed July 17, 2012.

[4] On teenage fears of nuclear war in the 1980s, see Robert T. Schatz and Susan T. Fiske, "International Reactions to the Threat of Nuclear War: The Rise and Fall of Concern in the Eighties" in *Political Psychology* 13, no. 1 (March 1992), 1-29; Paul Boyer *By the Bomb's Early Light* (New York: Pantheon Books, 1985), 352-368; see also William Knoblauch, "Selling the Second Cold War: Antinuclear Cultural Activism and Reagan Era Foreign Policy" (Dissertation, Ohio University, 2012).

[5] In the arcade era, the best examples of shooters include *Space Invaders*, *Galaga*, and *Robotron*; the most notable exception, of course, is *Pac-Man*.

[6] On atomic culture, see Boyer *By the Bomb's Early Light*; Scott Zeman and Michael Amundson, eds., *Atomic Culture: How We Learned to Stop Worrying and Love the Bomb* (Boulder: University of Colorado Press, 2004); Allan Winkler, *Life*

Under a Cloud: American Anxiety about the Atom (Urbana: University of Illinois Press, 1999); Spencer Weart, *The Rise of Nuclear Fear* (Cambridge, Mass: Harvard University Press, 2012).

[7] Harold Goldberg, *All Your Base Are Belong to Us: How Fifty Years of Videogames Conquered Pop Culture* (New York: Three Rivers Press, 2011), xi.

[8] The first-person shooter (FPS) genre has dominated action games since the mid-1990s. These are games in which the game is played through the protagonist's point-of-view. Examples include *Wolfenstein 3D, Doom, Duke Nukem 3D, the Call of Duty Series, Quake,* and *Fallout 3*.

[9] The much-lauded *Metal Gear Series* set new standards for stealth games. *Metal Gear* was first released in Japan in 1987, with an American release the following year. Some popular sequels include *Metal Gear 2: Solid Snake* (1990); *Metal Gear Solid* (1998); *Metal Gear Solid 2: Sons of Liberty* (2001); *Metal Gear Solid 2: Substance* (2003); and *Metal Gear Solid: Peace Walker* (2010).

[10] Years are provided in text for the first year in which a game was introduced. Video games, however, frequently go through numerous iterations and conversions to other platforms. I have only provided alternate years when new examples of atomic culture are present. Game producers are cited in the bibliography.

[11] Halter, *From Sun Tzu to Xbox*, 77, 85; Goldberg, *All Your Base Are Belong to Us*, xix-xxii; Rusel DeMaria and Johnny L. Wilson, *High Score: the Illustrated History of Electronic Games* (Berkeley: McGraw Hill-Osborne, 2002), 10-11.

[12] Lawrence Wittner, *Confronting the Bomb: A Short History of the World Nuclear Disarmament Movement* (Stanford: Stanford University Press, 2009), 141-176.

[13] Richards Rhodes, *Arsenals of Folly: The Making of the Nuclear Arms Race* (New York: Alfred A. Knopf, 2007), 114-117, 138-153.

[14] *Missile Command's* early title suggestions included "Armageddon" and "Ground Zero." The choice of terms was deliberate for a game with such a bleak theme.

[15] "Missile Command," in Rusel DeMaria and Johnny L. Wilson, *High Score!*, 60; "Coin-Op Capers: #10 Missile Command," *Retro Gamer* (Issue 88), 26-27; Theurer interview from "The Making of Missile Command," *Retro Gamer* (Issue 88), 62-65.

[16] SDI's roots are uncertain, but in 1940 Reagan starred in *Murder in the Air*, a film featuring a top secret device called the "Inertia Projector" which could fire highly concentrated particle beams to provide the "greatest hope for peace" in the world. Frances FitzGerald, *Way Out There in the Blue: Reagan, Star Wars, and the End of the Cold War* (New York: Simon & Schuster, 2000), 19-41.

[17] For a *Wargames* demo see http://www.youtube.com/watch?v=V18gkmSInaM, accessed May 20, 2012.

[18] Sega changed the game's title from *SDI* to *Global Defense* for its console conversion. On *High Frontier*, see "High Frontier," *ZZAP! 64* (November 1987), 123; for a demo of *SDI*, see http://www.youtube.com/watch?v=t+MTENzkSOg, accessed May 21, 2012.

[19] Gil Troy, *Morning in America: How Ronald Reagan Invented the 1980s* (Princeton: Princeton University Press, 2005), 235-264, esp. pp. 240-242; Sirota, *Back to Our Future*, 105-170; for a summary of American military adventurism

during the 1980s, see George Herring, *From Colony to Superpower: U.S. Foreign Relations Since 1776* (New York: Oxford University Press, 2011), 861-916.

[20] For *Raid Over Moscow* gameplay, see http://www.youtube.com/watch?v=33CCXg8B4aM, accessed April 20, 2012.

[21] Amundson and Zeman, *Atomic Culture*, 1-7.

[22] The *X-Men* comic books were first introduced in 1963, but have proved to be surprisingly durable, with a film franchise that started in 2000; *Godzilla* is an equally resilient piece of pop culture, with the radioactive lizard first appearing in 1954, but starring in numerous monster films since; *Them* appeared in 1954, but was simply one of many films inaccurately depicting radioactivity as having monstrous mutative effects. See Amundson and Zeman, *Atomic Culture*, 1-7.

[23] Ibid.

[24] A. Constandina Titus, "The Mushroom Cloud as Kitsch," in ed. Zeman and Amundson, *Atomic Culture*, 101-124; Knoblauch, "Selling the Second Cold War," 236-244; Lawrence Badash, *A Nuclear Winter's Tale: Science and Politics in the 1980s* (Cambridge: MIT Press, 2009).

[25] *Duke Nukem: Nuclear Winter*, WizardWorks (New York: GT Interactive, 1997), PC Game; CTF Map, http://www.tbhccs.com/treadmarks/ctf/ctf-index-low.html, accessed July 17, 2012.

[26] On atomic films, see Kim Newman's *Apocalypse Movies: End of the World Cinema* (New York: St. Martin's Griffin, 2000); Jerome Shapiro, *Atomic Bomb Cinema: The Apocalyptic Imagination on Film* (New York: Routledge, 2002).

[27] On survivalist literature, see "Rethinking the Unthinkable," *New York Times*, March 15, 1981, Sec. 6, pg. 45; Gladis Smith, "Reviews," *The Bulletin of the Atomic Scientists* (June/July 1983): 29-30.

[28] On the influence of the frontier thesis in culture and history, see Richard White, Patricia Nelson Limerick, and James R. Grossman, eds., *The Frontier in American Culture* (Berkeley: University of California Press, 1994); William Cronon, George Miles, and Jay Gitlin, eds., *Under an Open Sky: Rethinking America's Western Past* (New York: W.W. Norton & Company, 1992); *Wasteland's* cover art promised "adventure in post-nuclear America." See Imagine Games Network "Top 25 PC Games of All Time," online at: http://uk.pc.ign.com/articles/082/08221701.html, accessed May 30, 2012; Spencer Weart, *The Rise of Nuclear Fear* (Cambridge, Mass: Harvard University Press, 2012), 127-131.

[29] Tony Dillan, "Burntime," *CU Amiga*, 73; online at http://amr.abime.net/review_16842, accessed June 22, 2012; "Burntime," *Amiga Power* (August 1994): 40-41, online at: http://amr.abime.net/review_1597, accessed June 22, 2012; *Burntime* manual available online at http://www.oldgames.sk/en/game/burntime, accessed July 14, 2012.

[30] *Fallout's* implicit criticisms of 1950s Civil Defense propaganda are reminiscent of the 1982 film collection *The Atomic Café*; "The Vault Dweller's Survival Guide," a.k.a. the *Fallout* manual, included in the PC version of *Fallout* (Interplay Productions, 1997), 1-3.

[31] On the shift from radioactive mutation towards genetic mutation, see Amundson and Zeman, *Atomic Culture*, 1-7; Spencer Weart, *The Rise of Nuclear Fear*, 251.

[32] Ibid.
[33] Joseph Cirincione, *Bomb Scare: The History and Future of Nuclear Weapons* (New York: Columbia University Press, 2008), esp. 125-139; David E. Hoffman, *The Dead Hand: The Untold Story of the Cold War Arms Race and Its Dangerous Legacy* (New York: Anchor Books, 2009), 379-400.
[34] Zeman and Amundson, *Atomic Culture*, 1-9; Weart, *The Rise of Nuclear Fear*, 257-259.
[35] Detailed explanations or theories regarding the causes, and likely legacies, of American military involvement in the Middle East since 1991 remain outside of the scope of this essay. Still, a few useful sources on the subject include George Packer, *The Assassins' Gate: America in Iraq* (New York: Farrar, Straus and Giroux, 2006); Terry H. Anderson, *Bush's Wars* (New York: Oxford University Press, 2011); George Herring, *From Colony to Superpower*, 917-964.
[36] Writer Tom Bissell has called *Fallout 3* "George Jetson beyond Thunderdome." It's an apt summary. See Tom Bissell, *Extra Lives: Why Video Games Matter* (New York: Random House, 2010), 3-16, esp. p. 7.
[37] Sirota, *Back to Our Future*; On U.S. military uses of video games, see Halter, *From Sun Tzu to Xbox: War and Video Games*, 117-174.
[38] Paul Harris, "Bush Plans New Nuclear Weapons: 'Bunker-Buster' Bombs Set to End 10-Year Research Ban." *The Observer*, November 29, 2003, found online at http://www.guardian.co.uk/world/2003/nov/30/usa.georgebush, accessed July 17, 2012.
[39] Ibid.
[40] Note: There was another *Wargames*-based title. *Wargames: DEFCON 1* (1998) did appropriate the film's defense computer "WOPR" as a villain, but this game did not involve nuclear themes. Interview about *DEFCON* online at http://pc.vgcore.com/interviews/52.html, accessed July 17, 2012; *DEFCON's* webpage: http://www.everybody-dies.com, accessed July 18, 2012.
[41] For recent reflections on video games' importance, see Jane McGonigal, *Reality is Broken: Why Games Make Us Better and How They Can Change the World* (New York: Penguin Books, 2011); James Paul Gee, *What Video Games Have to Teach Us About Learning and Literacy* (New York: Palgrave MacMillan, 2007); Ian Bogost, *Persuasive Games: The Expressive Power of Videogames* (Cambridge, MA: MIT Press, 2010); Bissell, *Extra Lives: Why Video Games Matter*; Halter, *From Sun Tzu to Xbox*; levels of the nuclear threat are always changing. Three resources for up to date information include *The Ploughshares Fund*, http://www.ploughshares.org; *Project Ploughshares*, http://www.ploughshares.ca; and *The Bulletin of the Atomic Scientists* http://www.thebulletin.org/, all accessed July 18, 2012.

Chapter Seven

Depictions of Destruction: Post-Cold War Literary Representations of Storytelling and Survival in the Nuclear Era

Julie Williams

In his essay "No Apocalypse, Not Now," Jacques Derrida claims that in the absence of nuclear war, the only way society can deal with the presence of nuclear weapons is to write about them. This claim rests on the assumption that the use of nuclear weapons in Hiroshima and Nagasaki was the end of a "classical" war, not the beginning of a nuclear one. Within this framework nuclear conflict "can only be the signified referent, never the real referent (present or past) of a discourse or text."[1] The terrifying possibility of nuclear war is limited to literary representations, one of the most pervasive methods society used to deal with anxieties and fears in the Cold War era. The end of the Cold War did not bring an end to literature concerned with nuclear devastation; however, the methods authors use to predict possible apocalyptic futures and remember the sites of past nuclear tragedies have shifted in post-Cold War representations.

The historical transition from Cold War apprehensions is reflected culturally in the second phase of nuclear literature and Nuclear Criticism; this transition from Cold War to post-Cold War nuclear literature and criticism is explained by Daniel Cordle in *States of Suspense* as a "liberation" from representations of nuclear war itself (see Cordle's contribution to this volume).[2] Without the anxiety of the ever-present possibility of nuclear attack that shaped Cold War nuclear literature and criticism, post-Cold War nuclear literature and criticism are free to look at how nuclear motifs have impacted American culture in ways more subtle than the fictions of nuclear war would suggest. With his thoughtful analysis of this transition, Cordle lays the groundwork for a new Nuclear Criticism that focuses on representations of nuclear issues within society rather than the imminence of nuclear war, and his assessment of nuclear literature through the lens of anxiety provides the opportunity to explore the presence of nuclear issues in more mainstream texts.[3] Nuclear Criticism in the post-Cold War era should look to the ongoing representations of nuclear issues in discourse to understand the cultural attempt to conceive a future beyond the limits of Cold War anxiety, as well as how nuclear motifs still have resonances in our society despite political and historical changes. The end of the Cold War did not bring an end to the issues of nuclear waste, the legacy of the first and only use of atomic weapons in Hiroshima and Nagasaki, or the fear of the power of nuclear weapons still present in our world. The continuance of nuclear themes in post-Cold War writing indicates that the issues this literature addresses have not been resolved, and analysis of these ongoing representations opens up new ground for Nuclear Criticism to address both the material effects of the use of nuclear weapons as well as their psychological repercussions.

My main concern in this essay is how contemporary nuclear literature deals with depictions of the aftermaths of past nuclear events and imagined possible future catastrophes. Through my analysis of three post-Cold War literary texts which deal with the nuclear referent in very different ways, I will address the cultural implications of how we remember our nuclear past and the anxieties present in literary predictions about possible futures for a world where the nuclear question has not gone away. The ways in which Hiroshima's nuclear past are represented in Gerald Vizenor's *Hiroshima Bugi: Atomu 57* differ greatly from the depictions of a future apocalypse in Cormac McCarthy's *The Road*, and both are radically different from Terry Tempest Williams' *Refuge*, which looks at the consequences of nuclear testing in the American West.[4] These differences reveal much about how we as a culture want to remember the past and envision the future in addition to challenging what we think of as nuclear tragedy. These texts are not meant to be representative of all post-Cold War nuclear literature, but rather to serve as specific examples of how contemporary authors address ongoing nuclear issues. The continued resonance of nuclear imaginings in this literature demonstrates that the issue has not been silenced or forgotten, but that representations have shifted, reflecting the enduring cultural legacy of nuclear threats in a post-Cold War world.

While these texts represent nuclear issues in vastly different ways, they also share common themes. Each text in this study is centered on what happens after nuclear events rather than the event itself. As a part of this concern, they address notions of futurity and survival, aspects that have been essential to literature and criticism concerned with nuclear issues since the advent of atomic power. The issue of survival is present in the majority of literature concerned with nuclear themes, yet the idea of "survivance"—an active resistance to ideologies that threaten to destroy as much culturally as the havoc wreaked by forces like the atomic bomb destroyed physically—is unique. The role of stories factors greatly into the idea of survivance, as it requires practitioners to create their own narratives rather than accept the ones provided for them by governments, culture, religion, and other ideological forces. Each text in this study is occupied with the cultural values transmitted through storytelling, the types of stories being told, and the manner in which they are transmitted. These concerns link the texts I address despite their disparate representations of nuclear issues, and the common themes bring to light the continued relevance of Nuclear Criticism in the post-Cold War era. Each text deals with the reality of life in the nuclear age that Derrida distinguished from the fable of nuclear war. A critical aspect which links the texts in this study is the way in which the nuclear events exist at the periphery of the narratives yet at the same time have a tremendous impact on the trajectory of the plot. This aspect reflects how the nuclear referent has moved to the periphery of our cultural consciousness but has not yet gone away.

The way in which these texts represent nuclear events as peripheral to the main narrative structure reflects a significant change from Cold War-era nuclear literature focused on anxiety and disaster. This corresponds to Derrida's open definition of nuclear literature, where he claims that "[l]iterature has always belonged to the nuclear epoch, even if it does not talk 'seriously' about it. And in truth I believe the nuclear epoch is dealt with more 'seriously' in texts by Mallarme, of Kafka, or Joyce, for example, than in present-day novels that would offer direct and realistic descriptions of a 'real' nuclear catastrophe."[5] The texts in this study do not offer direct descriptions of nuclear catastrophe, yet concerns associated with the use (or possible future use) of nuclear weapons and the humanity that possesses them lurk in the background of each text. This signals a shift from what Cordle calls "disaster narratives," Cold War era nuclear literature which blatantly deals with nuclear fear and the potential for disaster, as there were "many more narrative possibilities in the event itself" than in the "definitively non-eventful dimensions of nuclear anxiety."[6]

With the tensions of the Cold War behind us, the nuclear threat has not disappeared from the cultural consciousness. Literature in the post-Cold War era has proven that the "non-eventful dimensions" provide a serious way to represent this continued threat, and the ways in which the three narratives in this study examine the aftermaths of nuclear events provide an insightful look into the reality of life in the nuclear age. In examining the nuances of how nuclear weapons have altered our world, rather than depicting the event of nuclear war, this literature continues to push the boundaries of how we deal with the nuclear referent in our society. Such an analysis opens up space for a new Nuclear Criticism, for Nuclear Criticism has, from its early articulation in Derrida's inaugural essay, "thought about the limits of experience."[7] The literature examined in this essay challenges previous notions of the use of nuclear weapons and the legacies of nuclear tragedies as it depicts a world in which the power of the split atom remains.

The Legacy of Nuclear Peace: Gerald Vizenor's *Hiroshima Bugi: Atomu 57*

Gerald Vizenor, widely known for his contributions to Native American literature and theory, published *Hiroshima Bugi: Atomu 57* in 2003 in response to the sentimental presentation of Hiroshima as a "city of peace and victimry."[8] The novel, which contains elements of both Anishinaabe and Japanese traditions and has narrative voices located in both Hiroshima and Nogales, Arizona, resists easy classification as atomic-bomb literature, Native American literature, or a postmodern tale. However, the method with which *Hiroshima Bugi* takes on the politics of the peace movement and its concern with the stories told in remembrance of this movement makes it a text that is rich with meaning and a valuable source for the study of contemporary nuclear literature. Vizenor's insistence upon privileging the ongoing effect of the American occupation and bombing of Japan is a shift from the previous phase of nuclear and atomic bomb literature, which focused primarily on the events of the war rather than the repercussions that those events continue to have in both American and Japanese society.

Hiroshima Bugi challenges the boundaries between atomic-bomb literature and nuclear literature. By questioning the ways in which the aftermath of the bombing of Hiroshima was dealt with in Japan, the novel refuses to view this event in isolation. Instead, *Hiroshima Bugi* examines the role of Japanese aggression prior to and during WWII as well as the effect of nuclear testing in the United States.[9] The narrative centers on the main character's battle against the "fakery" present in the Hiroshima Peace Memorial, and as such it deals with a theme central to atomic-bomb literature: the effects of the nuclear bomb on Hiroshima. However, the text is also concerned with themes that are not typical of this genre: the identity of Ronin, the main character, his ties to the Anishinaabe tribe of which his father was a member, and the native skills of irony, storytelling, survivance, and wit that he has inherited from this lineage and puts to use in post-war Japan. The story has dual narrators, Ronin Browne, a half-Japanese and half-Anishinaabe orphan who now lives in the Atomic Bomb Dome, and an anonymous best friend of Nightbreaker, Ronin's father, who resides at a home for wounded native veterans in Nogales, Arizona. Vizenor's narrative focuses on how people attempt to understand the atrocity and how its meaning was co-opted in the ensuing move to memorialize the event. The concern with matters of conscience and how this event is remembered are evident from the beginning of this text, and point to its position within a tradition of nuclear literature which shares these concerns.[10]

The novel begins with Ronin declaring that "The Atomic Bomb Dome is my Rashomon."[11] From the start, readers are invited to recognize the subjective nature of truth and the complexities of human nature with the reference to Rashomon. Both the Rashomon of Akira Kurosawa's 1951 movie and that of Ryunosuke Akutagawa's 1915 short story of the same name resonates with Ronin's declaration, as they are each concerned with moral ambiguities seen as necessary to survive.[12] Akutagawa's short story shares similarities with the setting of Ronin's Rashomon in *Hiroshima Bugi*, as the short story takes place in a dilapidated gate in the devastated city of Kyoto. This setting correlates with the Genbaku Dome in Hiroshima, which remained partially intact despite being almost directly under the detonation of the Little Boy bomb dropped on August 6, 1945. After the bombing, the building almost certainly saw scenes to rival the gruesome one told in the short story of an old woman stealing hair from the corpses that were dumped in the dome in order to make wigs to sell for the living. The gate in the Kurosawa film thus provides a setting for a similarly ambiguous portrayal of humanity; however, the most striking congruency between Kurosawa's and Vizenor's Rashomon is the focus on the contradictory nature of multiple testimonies and the uncertainty of truth when dealing with memories of an event. Ronin declares that we cannot read his story as an absolute truth, and the readers of this text are called upon to consider the ways in which all stories, particularly those as complex and emotional as ones told about nuclear tragedies, reveal the subjectivity of memory and truth.

The connection to Kurosawa's allegory about the subjective nature of truth is heightened through the presentation of Ronin's story, in both his own words and through the notes of the unnamed narrator. Ronin meets this friend of his father's when he tracks down his missing father at the "Hotel Manidoo," a "hotel of perfect memories for wounded veterans" in Nogales, Arizona.[13] He arrives too late to find his father, Orion Browne (or Nightbreaker), who died the week before. But he learns much about his father's history from the stories related by his father's friend. Vizenor plays with the connection to Kurosawa's film in this introduction of the two narrative voices, as the unnamed narrator declares that Ronin "could have been mistaken for Toshiro Mifune. He bounded into the room as if he were on the set of the movie *Rashomon*."[14] The readers are told that Ronin eventually entrusted his journal to this narrator, with the instructions to add notes and background information before publishing the story, and this explains the narrative construction of the novel: following each chapter of Ronin's is a "Manidoo Envoy," which contains the veteran's explanatory notes and commentary. The addition leads to a fractured narrative portrayal of this singular experience, an apt representation of the fragmentation of life in post-war Hiroshima. Readers also learn that they are not the first to hear this story, as "*Hiroshima Bugi* was read out loud at dinner by the storiers named for the day. Ronin would be pleased to hear the creative counts that became part of his tricky stories, and, of course, my commentaries."[15] This statement reveals even further narrative perspectives, as Ronin's story is not only influenced by the commentaries of the other narrator, but also by the "creative counts" of other storiers, the native veterans who were the residents of Hotel Manidoo.

The role of "storiers" is often emphasized in Vizenor's novels and theory, as he recognizes the power of stories to shape the reality they describe. In *Hiroshima Bugi*, Ronin sees the importance of creating his own stories about the Atomic Bomb Dome, for if he declines to tell his truth, the stories perpetuated by the "simulations of peace" presented in the Peace Memorial Museum will not be countered.[16] He learns from the residents of the Hotel Manidoo to craft stories that resist the indoctrination perpetrated by those who dominate the dialogue, to resist victimry and instead create an active presence for an "atomu hafu" (atomic mongrel) such as himself and the other roamers who live with him in the Atomic Bomb Dome: Oshima the leper, Virga his canine companion, and Kitsusuki the wounded veteran. The residents of Hotel Manidoo too are well versed in creating stories that resist a sense of victimry, as the veteran narrator emphasizes when describing their routines at the Hotel, saying that "Five nights a week we came together for dinner and to create our perfect memories. The marvelous, elusive tease of our many stories, became concerted memories. Our tricky metaphors were woven together day by day into a consciousness of moral survivance."[17] The use of metaphor and the role of a creative group construction of memory are essential for understanding Ronin's own sense of storytelling, as is the focus on "survivance" as opposed to survival, as "survivance is a creative, concerted consciousness that does not arise from separation, dominance, or concession nightmares."[18] Ronin employs similar narrative strategies in constructing the text that would become *Hiroshima Bugi*. He weaves metaphors of life in Japan into the creative story of nuclear survivance that is meant to counter the reductive stories that turn *hibakusha* (explosion-affected persons) into victims, deny the role of Japanese aggression in WWII, and use simulations of peace to avoid the real issues at stake in the nuclear era.

Though the importance of stories is not unique to *Hiroshima Bugi*, the way in which stories of victimry or survival and acceptance or resistance are told about past, present, and possible future nuclear events reveals much about how a society intends to document their relationship to nuclear issues. As the contradictions and limitations of the few largely unquestioned discourses of the occupation government became apparent, people who experienced the bombings and their after-effects felt an urgency to create a textual representation that more accurately represented their experiences. The atomic-bomb literature published in the post-war years was rarely written by professional writers; rather, it was created by *hibakusha* who felt an urgency to share their experiences with the world so that others might know of the horrors associated with the nuclear blasts. In *Fire From the Ashes: Short stories about Hiroshima and Nagasaki* (1985), Oe Kenzaburo collected post-war literature which wrestles with understanding the damage caused by the bombings, as well as the "second-generation" literature which more openly questions the responsibility of the Japanese government and their part in WWII and the Fifteen Years War leading up to the bombings. The stories collected in *Fire From the Ashes* seem cognizant of their role in relating a Japanese perspective on the larger political and ethical issues associated with the use of nuclear weapons. Oe describes the importance of telling these stories, including those that veer from the confines of the official, government-censored narratives of occupied Japan, in his introduction:

> I have come to realize anew that the short stories included herein are not merely literary expressions, composed by looking back at the past, of what happened at Hiroshima and Nagasaki in the summer of 1945. They are also highly significant vehicles for thinking about the contemporary world over which hangs the awesome threat of vastly expanded nuclear arsenals. They are, that is, a means for stirring our imaginative powers to consider the

fundamental conditions of human existence; they are relevant to the present and to our movement towards all tomorrows.[19]

Ronin Browne, for his part, recognizes that the importance of these types of stories did not diminish with the passing of time after the atomic bombings, and he takes up the challenge of creating modern day expressions for considering the "conditions of human existence." The desire to relate his experience as an atomic orphan can be compared to how survivors of the bombings in Hiroshima and Nagasaki felt an urgency to tell their stories, as they all feel the need to exert their existence in a country which would often prefer to ignore them or to use their plight for political purposes. The importance of these stories connects the texts included in this study, as the stories we tell, and how they are told, disclose how we think about the modern society which contains—and has used—nuclear weapons. These stories manifest core elements of humanity through their examination of how we manage to go on living in a world which maintains the horror of nuclear weapons; they transmit cultural values through their challenge of official narratives.

By telling his stories using irony and creative wit, Ronin is countering the narratives that are conveyed in the Peace Memorial Museum, a place he sees as a "cynical theme park of human misery," and the letters written by worldwide leaders and peace activists and etched into a metal column in the museum that he considers "a testament to the arrogance and deceptions of political peacemongers."[20] The museum is a simulation of the real Atomic Bomb Dome, built because many residents of Hiroshima were critical of the real ruin due to the reminder it provided of the devastation of the bombing. Ronin despises the way that the museum, a tourist attraction, presents Japanese history, the aftermath of the bombing, and the plight of the *hibakusha* to visitors. Although he admits that "some letters seem to be sincere," it is the ideology behind their presentation in the museum that he cannot abide, as "[t]he museum elevates the peace letters, the government solicits a free ride on the passive road to peace, and, at the same time, there are tricky moves to contract nuclear weapons in the country."[21] He destroys the letters in a scene influenced by the Japanese tradition of Kabuki Theater, with the museum guests too busy in the gift shop to notice the scene of destruction. After throwing a corrosive chemical onto the metal etchings, the letters "rightly vanished forever in a wispy cloud of silence."[22] This silencing of the official narrative of the bombing of Hiroshima, and the meaning it carries in current political circles, can be seen as revenge against the Japanese government, for we learn from the veteran narrator in the following chapter that it was eager to portray its country as victims of the Americans, as "only a few untitled photographs in the museum represent the atrocities of the Japanese military in Shanghai and other cities in China."[23] Additionally, readers learn that this type of government-censored official narrative is not new, as the narrator quotes Ota Yoko's *City of Corpses*,[24] saying that "[t]hroughout the war, we could not be true to our own selves [...] We lamented the fact that we could not say what we wanted to say; but also had to say and do things we didn't want to say and do."[25]

Ronin views the presentation of these skewed stories as a disservice not only to the visitors of the museum, but also to the *hibakusha* themselves. By presenting their country as victims, the Japanese government silences the stories of the survivors of the atomic blast and diminishes their vision of themselves as enduring despite drastic emotional, monetary, and physical challenges. Challenging this route of victimry, Ronin views atomic orphans such as himself and others affected by the bombing as taking the course of survivance, which is "not merely a variation of 'survival,' the act, reaction or custom of a survivalist. By 'survivance' he means a vision and viral condition to endure, to outwit evil and dominance, and to deny victimry."[26] This idea of survivance is unique, even though the issue of survival is present in many nuclear texts. The role of stories is crucial to the idea of survivance, as it requires the creation of new narratives rather than acceptance of the ones given to us by ideological forces.

Ronin creates a presence for himself that defies the ideologies of both Japan and the United States, as both countries refused to embrace the children of American soldiers and Japanese women and saw them as "the untouchables of war and peace in two countries."[27] In reality, these children were viewed as racially inferior in Japan. At the same time, the United States enacted restrictive immigration laws which made it difficult for such children to be adopted. In the absence of his father, the White Earth Reservation adopts Ronin, asserting a sovereign right beyond the restrictions of U.S. immigration policy. The veteran narrator relates that Ronin "has died many times over the manners and proprieties of an empire nation that would not embrace the *hafu* children of the occupation," a description that applies as easily to the United States as to Japan.[28] Yet death for Ronin is the "bushido way" as opposed to a defeat, and at the end of the novel when his lover Miko visits White Earth Reservation hoping to honor his memory, "she was teased at the mere mention of his name. Ronin was known on the White Earth Reservation, to be sure, but not the way she constructed the story."[29] He resists stable interpretation to the very end, as "the Rashomon effect" carries through the entire narrative.

Ronin's father, first exposed to radiation in occupied Hiroshima, was again exposed during atomic testing at the Nevada Test Site, and the combination led to a cancer diagnosis and his retirement from the army to "nurse his nuclear wounds."[30] The veteran narrator describes his and other military veterans' experience as participants in atomic tests at Yucca Flats, Nevada, relating the sunburn and nausea they felt immediately after the test to the tooth and hair loss that eventually ensued after the nuclear blast, saying that "The Nevada Test Site was their Rashomon."[31] This description of Nightbreaker's experience in the Nevada desert opens up a whole new set of nuclear problems that remain unanswered in *Hiroshima Bugi*, yet the set of stories that deal with atomic testing is just as rife with moral ambiguities and the need to counter the official government narratives as Ronin's narrative. The experience of soldiers and downwinders differs from the *hibakusha* of Hiroshima and Nagasaki, as the nuclear tests were not intended to devastate a city and its population in the way that the bombs dropped on Hiroshima and Nagasaki were. Yet these areas and people are connected through their exposure to nuclear radiation. Problems from the radiation exposure challenge what we consider to be nuclear tragedies and how we conceptualize the use of nuclear weapons. The reality for those exposed to nuclear blasts does not consist of an anxiety about the possibility of nuclear tragedy, but that of the effects from it. As with all of the texts in this study, this reality depicts the anxieties that Cordle sees as characteristic of post-Cold War nuclear literature. This theme plays a large part in Terry Tempest Williams' *Refuge*, and so it is to this text that I now turn.

Pushing the Boundaries of Nuclear Tragedies: Terry Tempest Williams' *Refuge: An Unnatural History of Family and Place*

Terry Tempest Williams' *Refuge: An Unnatural History of Family and Place* has become a landmark of modern nature writing since its publication in 1991. Williams' memoir relates the personal tragedy of her mother's cancer diagnosis alongside the community concerns about how the rising levels of the Great Salt Lake threaten the Bear River Migratory Bird Refuge and Salt Lake City itself. The text combines these issues in a narrative which strives to find the human place in nature's design despite the difficulties humanity has caused to the natural world through atomic testing, overpopulation and overdevelopment, and political mismanagement of resources and land. Just as Vizenor's *Hiroshima Bugi* resists classification into a single genre of literature, I believe that *Refuge* has as much to offer the field of Nuclear Criticism as it has provided to ecocriticism, and that as Nuclear Criticism progresses in the post-Cold War era it would do well to take note of the issues that ecocriticism addresses. Analyzing *Refuge* through the lens of Nuclear Criticism highlights one way this field can serve as a useful method for examining life in a post-Cold War world, as the issue of nuclear testing and the fallout to which people around the test sites were exposed was not resolved with the end of the Cold War and continues to affect residents of Nevada and Utah and the landscapes in which they live. Additionally, *Refuge* takes up the themes of survival and storytelling that were central to *Hiroshima Bugi*, further emphasizing the importance of these issues. While Williams' framing of the issue of survival is, on the surface, quite different than the way that Vizenor addresses this issue in *Hiroshima Bugi*, in both texts characters refuse to accept dominant beliefs about the issues of victimry and survival, and both of these texts narrate "survivance" through the characters' assertion of their own creative expression.

The issue of atomic testing pushes to the forefront the ways in which nuclear weapons were used during the Cold War, both in the American West and the Marshall Islands. While Nuclear Criticism writes of the fable of nuclear war, the reality of nuclear weapons has proven to be quite different than critics anticipated. The reality of nuclear weaponry differs from that of all-out nuclear war, yet anxieties about the possibility of nuclear war and the Cold War policy of deterrence led to the atmospheric atomic tests that were conducted in the Nevada desert during the 1950s and early 60s. The fallout from these tests became the Cold War reality of "downwinders," the name for residents of Nevada and Utah who were in a direct line of exposure to the radiation from these tests, and the consequences have been devastating to the health of those affected. In a time when the nuclear referent has become diffused in literary and cultural imaginings, the issue of fallout from nuclear testing remains an area where Nuclear Criticism can turn, as "the 'reality' of the nuclear age and the fable of nuclear war are perhaps distinct, but they are not two separate things."[32] The issue of nuclear testing makes this connection painfully clear.

Refuge shares a sense of indeterminacy with *Hiroshima Bugi*, although Williams portrays this through the shifting of natural rhythms in the landscape surrounding the Great Salt Lake rather than a fragmented narrative style. Though the narrative style differs, the effects are the same: readers are moved through the text in a way that resists a stable, definitive interpretation. The Prologue begins by describing the Great Salt Lake as "a landscape so surreal one can never know what it is for certain."[33] The changes in the landscape become a metaphor for the changes in her family as her mother and other female family members are diagnosed with cancer, and the uncertain nature of the land and water surrounding her give Williams comfort as she searches for a sense of understanding about why this is happening. The uncertainties and shifts of nature provide a way for Williams to comprehend the tragedy her family faces. The narrative structure reflects the way in which the natural world provides a frame through which she can put her personal difficulties into perspective.

Each chapter is named after a bird that Williams has seen at the Great Bear Migratory Bird Refuge, a place which becomes as much a refuge for her as for the birds she watches there. The level of the Great Salt Lake is also recorded at the beginning of each chapter, which provides readers a concrete way to understand the amount of change that this landscape is undergoing. This narrative structure emphasizes both the importance of nature and the way in which it is in constant change, an emphasis which seems necessary for both Williams and her mother Diane when confronted with cancer for the second time in her life. Both mother and daughter turn to nature in order to understand and accept the sickness. After finding a mass in her abdomen, Diane takes a river trip down the Grand Canyon before she visits her doctor, saying that "the Grand Canyon is a perfect place to heal" and "I needed time to live with it, to think about it—and more than anything else, I wanted to float down the Colorado River."[34] The comfort she finds in nature, and the desire to be out in it, surpasses even the caretaking of her body. When faced with chemotherapy, Diane asks Williams to "help me visualize a river—I can imagine the chemotherapy to be a river running through me, flushing the cancer cells out."[35]

When processing her mother's illness, Williams attempts to rethink the metaphors associated with cancer, preferring to think of the process of cell division that eventually leads to a cancerous tumor as akin to the creative process. She resists thinking of illness as something to conquer, and refuses to accept the way in which "medical language is loaded, this time with military metaphors: the fight, the battle, enemy infiltration, and defense strategies."[36] For someone whose main source of comfort is the natural world, these military metaphors provide little comfort, and her meditations on the subject lead her to wonder if "aggression waged against our own bodies is counterproductive to healing? Can we be at war with ourselves and still find peace?"[37] When viewed through the lens of Nuclear Criticism, her critique of the way that we think of disease in our culture takes on added meaning, as these questions themselves can be read as metaphors for how nuclear testing was being conducted in the desert that Williams loves. By testing nuclear weapons in an area where its own citizens were at risk, the United States was in a sense waging aggression against its own body politic; one could see this phenomenon as Americans being at war with themselves. Rebecca Solnit, in *Savage Dreams: A Journey into the Hidden Wars of the American West*, asserts that "[n]uclear war, whether you are for it or against it, is supposed to be a terrible thing that might happen someday, not something that has been going on all along."[38]

Solnit examines a myriad of issues surrounding the Nevada Test Site, from the land battles with the Western Shoshone and local ranchers to the ideologies of "manifest destiny" and "American exceptionalism" that led to the selection of the Test Site in an area that was "a blank on many maps, a forgotten landscape."[39] Her conception of the atomic testing that occurred at the Nevada Test Site asserts the reality of nuclear war waged in a very different way than the literature of the Cold War era anticipated. Solnit rejects even the word "test," instead asserting that the physicists and military personnel were "rehearsing the end of the world…over and over again."[40] Both Solnit and Williams are intensely concerned with landscape and politics of place, and the military aggression waged against the Nevada desert hits too close to home for Williams to engage in the use of military metaphors to describe her mother's illness (a form of cancer that may have been caused by fallout from the tests).

The tests themselves exist at the periphery of the narrative, even though underground testing was still happening at the time Williams wrote *Refuge*. The issue is not approached until midway through the text, but when the nuclear referent is finally addressed we see how it has played a part in Diane's feelings about her illness all along. After cancelling their planned hike in the Mojave Desert because "another nuclear bomb was being detonated underground at the Nevada Test Site," Williams, her mother, and her grandparents "seek quiet" in St. George, Utah when they encounter a group of protesters.[41] The "another" of this comment reveals that this is a process that has been occurring throughout the course of the narrative, despite the silence about the issue. They are all moved by the procession of protesters, who are described as "a slow-moving river, hundreds of people walking on behalf of nuclear disarmament. The Great Peace March."[42] The description of the protesters as a river is telling, as Diane had previously envisioned the chemotherapy that she hoped would cure her cancer as a river running through her, eliminating the cancer cells and healing her body. The use of the river metaphor to describe the protesters reveals her opinion on the issue of nuclear testing, placing the protesters as the river that will heal the desert from the "cancer" of the atomic detonations. The scene continues to affect them, particularly Diane, who sees the connections between her situation and the reasons behind the march: "'I could join them,' Mother said under her breath as we clapped for them. A song rose up from the activists: *We are a gentle, loving people and we are walking, walking for our lives*—We walked with them. It was the first time I had ever heard Mother and Mimi sing outside of church."[43] Diane's relationship to the song about "walking for our lives" is complicated by her views on how to accept illness and end-of-life issues, yet the shared moment of protest remains a pivotal point in the text.

The issue of survival focuses upon finding pleasure in day-to-day life experiences rather than trying to prolong the time one has. This perspective, easy for Diane to see from her vantage point, is hard for those around her to accept. Williams and the other family members continue to hope that she will "get better" throughout much of the narrative, an outlook which leaves Diane frustrated: "'You still don't understand, do you?' Mother said to me. 'It doesn't matter how much time I have left. All we have is now. I wish you could all accept that and let go of your projections. Just let me live so I can die.'"[44] This statement provides perspective on how she might have related to the line "walking for our lives" sung by the protesters during the anti-nuclear march. Rather than viewing the march as a means to ensure future life, Diane was likely to have viewed the walk as a part of her life, an event in which she participated and enjoyed because it connected her both to the desert that she loved and to the like-minded individuals that surrounded her.

The philosophy that Diane draws upon in this passage, that there is no retreat from or extension to life, and that "all we have is now," is often lost in the focus on prolonging the time available. Diane's desire to "live so I can die" is a departure from how both illness and dying are generally approached in the United States. Yet, just as she rejects the military metaphors of battling her disease and fighting her own body, she also rejects her family's hopes that she can beat her cancer. For those around her, this rejection is hard to accept. However, Diane's terminal diagnosis does not prevent her from living fully, even when that means embracing the process of dying. How Williams frames the issue of survival in the nuclear era is wildly different than how Vizenor takes up this theme in *Hiroshima Bugi*, but the way in which Diane refuses to accept the dominant beliefs about how one should experience illness and the process of dying has resonances with the way in which Ronin and the veterans at Hotel Manidoo refuse to accept the narratives of dominance, instead choosing to assert their own creative expression.[45] Throughout her illness, Diane declines to let the dominant American culture or the Mormon

Church dictate how she will relate to her own decaying body and approaching death. When asked by her daughter about what she believes, she replies that "I believe in me," reminding Williams that she refuses to define herself through anyone else's terms.[46]

Like the rest of the texts in this study, *Refuge* is concerned with the power of stories; the idea that stories shape our lives and systems of belief structures the text. Williams bookends the narrative by establishing which stories influenced her relationship both to herself and to the world around her. She sets up this dynamic in the prologue, writing that "[p]erhaps, I am telling this story in an attempt to heal myself, to confront what I do not know, to create a path for myself with the idea that 'memory is the only way home.' I have been in retreat. This story is my return."[47] The narrative expands on this process of confronting the unknown, as the entire memoir is composed around the theme of how to find refuge in the changes occurring in the natural world and in her family life. In the epilogue, "The Clan of One-Breasted Women," discussed at greater length by Dan Cordle in his contribution to this volume, both she and the readers realize the larger implications of how stories inform our view of the events around us. Williams therefore comes to understand "the deceit I had been living under" about the reality of nuclear testing and how it had affected her life.[48]

This realization occurs when she shares a recurring dream about a flash of light in the desert, a dream her father reveals to be true:

> "You did see it," he said. "Saw what?" "The bomb. The cloud [...] It was an hour or so before dawn, when this explosion went off. We not only heard it, but felt it. I thought the oil tanker in front of us had blown up. We pulled over and suddenly, rising from the desert floor, we saw it, clearly, this golden-stemmed cloud, the mushroom. The sky seemed to vibrate with an eerie pink glow. Within a few minutes, a light ash was raining on the car." I stared at my father. "I thought you knew that," he said. "It was a common occurrence in the fifties."[49]

The way that atomic testing was here revealed as an unknown presence in her life leads Williams to question how American society processes the issue of atomic testing and what stories we are told and those we tell ourselves about what and why and how this has occurred, questions which led her to reject the official narrative of the U.S. government: "It has been found that the tests may be conducted with adequate assurance of safety."[50] Although she cannot prove that the radioactive fallout caused the cancer in her mother, her grandmothers, and her aunts, she also cannot prove that it did not, and what that leaves her with is more questions: "The more I learn about what it means to be a 'downwinder,' the more questions I drown in."[51] This endless questioning is a very Derridean moment in the text, reinforcing the resistance of the narrative and the issues it raises against a stable, singular interpretation.

These questions ultimately remain unresolved. Williams realizes that everything must be questioned, as "[t]olerating blind obedience in the name of patriotism or religion ultimately takes our lives."[52] She ends the narrative with a story that reflects the way she hopes to live after realizing the deceit about how nuclear issues are handled in our society. She shares with her readers a dream about a group of women protesting the contamination in the desert. In the course of this story, she moves from speaking of "the women" to "we" to "I" as this dream, like the earlier one that revealed the hidden presence of nuclear testing in her life, proves to be true. The text leaves readers with the image of her crossing the line onto the Nevada Test Site in an act of civil disobedience, getting arrested, and then transported out into the desert, a joke on the part of the officers to leave the women stranded when in fact they are already home. The way that Williams addresses the issues of home, storytelling, and survival in *Refuge* reinforces the sense of indeterminacy that Vizenor's text creates, and these themes continue to be of importance in the dystopian possible future created by the next text in this study, Cormac McCarthy's *The Road*.

A Threatened Future: Cormac McCarthy's *The Road*

While Terry Tempest Williams concludes *Refuge* with the sense of being comfortably at home in one's environment, Cormac McCarthy's *The Road* provides the opposite: the journey of a father and son adrift in an apocalyptic future with no concrete destination and little hope that their wandering will ever stop. Unlike the two other texts included in this study, *The Road* does not deal with our nuclear past; rather, it imagines a desolate future in which disaster has struck and the characters are desperately clinging to life in a world which bears little resemblance to the one readers know. The manner with which the narrative presents life in the aftermath of this unnamed disaster signals it as particularly post-Cold War, as Cold War disaster narratives tend to focus upon the lead up to, or the experience of, the disaster itself. McCarthy's text is almost solely concerned with life a number of years after the event itself, yet the event remains unnamed and at the periphery of the narrative. These depictions both fit the political climates of their times, as the Cold War era dealt with the anxiety of what seemed like ever-impending nuclear disaster, while McCarthy's text, published in 2006, suggests a more diffuse set of tensions, reflecting how apocalyptic anxiety has expanded to include a post-9/11 fear of terrorism, the increasing concern of global warming, and in the past few years, the fear of economic collapse. McCarthy positions the disaster that led to the world depicted in *The Road* at the periphery of the text and so reminds us of the texts included previously in this study. All three of these narratives suggest the type of anxiety that Cordle pinpoints as the hallmark of post-Cold War nuclear literature: the texts are structured by the nuclear referent, yet the idea of nuclear disaster lurks in the background. The broken or disjointed narratives suggest the importance of locating personal truth in the face of ideological fictions, or in the case of *The Road*, of an overwhelming sense of meaninglessness.

The images used to describe the devastated world of *The Road* call upon the imagery associated with nuclear war and an ensuing nuclear winter.[53] As the father and son journey south hoping to find an enclave of people who still have a sense of humanity, they pass through a barren landscape, with scenes reminiscent of how we have been taught through popular discourse to imagine the aftermath of nuclear war. Passing through a city, they discover it to be "mostly burned. No sign of life. Cars in the street caked with ash, everything covered with ash and dust. Fossil tracks in the dried sludge. A corpse in the doorway dried to leather."[54] They come upon people long ago burned to death in the road as fires engulfed the entire area, just as fires destroyed many of the buildings that withstood the initial blast in Hiroshima. The image of black rain, also invoking the bombing of Hiroshima, is referenced throughout the text, along with a darkened sun and the ensuing drop in temperature. The event carries the markers of a nuclear bomb as the readers find out when the man reflects back on a night which seems to be the start of the disaster: "The clocks stopped at 1:17. A long shear of light and then a series of low concussions. He got up and went to the window. What is it? she said. He didnt answer. He went into the bathroom and threw the lightswitch but the power was already gone. A dull rose glow in the windowglass."[55] The flash of light and "series of low concussions" invoke the sights and sounds associated with nuclear weapons, even though they are never explicitly named, and the stopped clock and the immediate loss of electricity signal the passing into a new era in which light, both literally and symbolically, is scarce.

The woman with the main character at the time of the disaster was pregnant with their son, who is born in the bleak world that he and the nameless father now experience. She disappears from the story in the early pages, and the issue of survival present in all of these texts is brought to the forefront by her desire to commit suicide. Trying to talk her out of her plan, the man tells her they are survivors, a word she rejects: "We're not survivors. We're the walking dead in a horror film."[56] Survival for her seems to mean more than endless walking, scraping together barely enough food to stay alive, and living in constant fear of fellow humans who have lost their sense of humanity. Determined to escape the horror of the world they inhabit, the woman leaves the man with parting advice, saying that "[t]he one thing I can tell you is that you won't survive for yourself [...] A person who had noone would be well advised to cobble together some sort of passable ghost. Breathe it into being and coax it along with words of love."[57] This explains the man's fierce determination to survive, as he is the sole protector of their son. He tries to provide the boy with the skills necessary to carry on, even after his inevitable death.

His love for his son motivates all his actions, although at times it interferes with the sense of ethics he tries to instill. Born after the disaster, the son has no memory of the world which came before, and sometimes questions his father about how things used to be. The father is often conflicted about how to answer, as the contrast between their experience and life in the past seemed too harsh to relate. Yet he cannot stand to crush the hope the boy fosters, that things would be different at some other point in space or time, by allowing the bleakness of their world to be the only thing that the boy knows. As he explains, "There is no past. What would you like? But he stopped making things up because those things were not true either and the telling made him feel bad."[58] He struggles with the desire to paint a picture of a better world for his son when the circumstances they face seem to make this illusion cruel, and his actions often reveal the inconsistency between the reality of their world and what he tells the boy. When a thief steals their shopping cart, the container for their food and all the goods necessary for their survival, the father catches up with him and demands that he not only give back their belongings but also give them his clothes, the only thing the thief possesses. His son begs him to stop, but the father is adamant, and they leave the thief naked and shivering in the middle of the road. "I wasnt going to kill him, he said. But the boy didnt answer. They rolled themselves in the blankets and lay there in the dark. He thought he could hear the sea but perhaps it was just the wind. He could tell by his breathing that the boy was awake and after a while the boy said: But we did kill him."[59] The boy recognizes that in their world taking someone's clothes was enough to throw off the delicate balance between life and death. Influenced by the stories his father told about "carrying the fire," the boy becomes increasingly upset by the discrepancy between their actions and the way the father depicts their role in the world.

The distress the boy feels about the inconsistencies between their values (the stories they tell) and their actions brings to light the role that stories play in this text. In a world where survival is the highest priority there is often very little to separate the "good guys" from the "bad guys." The one issue that consistently separates the father and his son from those that they fear is cannibalism. The man knows that he will not be able to protect his son forever and tries to instill in him the skills necessary to be an ethical human being as well as those that will allow him to carry on and avoid getting captured by the "bad guys." Although they do not turn to cannibalism, the boy knows that their actions do not always make them good. They encounter a number of people along the road and the boy, who never had companions other than his father and mother, is eager to adopt the ones that seemed like them: hungry and bordering on desperation, but still "carrying the fire." Protective of their scarce resources, the father denies the boy's pleas every time, which is particularly painful when the boy sees another child around his age, as his fears about being left alone to fend for himself become evident. After this occurs a number of times, the boy stops wanting to engage in storytelling with his father: "Do you want to tell me a story? No. Why not? The boy looked at him and looked away. Why not? Those stories are not true. They dont have to be true. Theyre stories. Yes. But in the stories we're always helping people and we dont help people."[60] The boy recognizes the way that the stories do not match the reality in which they live, and is forced to question either the stories—and the values they teach—or the father's enactment of those values in the world. In *Hiroshima Bugi* and *Refuge*, characters question the narratives put forth by government powers; in contrast, when the government and other ideological apparatuses disappear in *The Road*, the father becomes the authority that needs to be scrutinized. This text solidifies the need to question the dominant narratives that are told in society, whether that society is post-war Japan, the American West, or the father and son in the unspecified ruins of *The Road*. In order to survive in a world where nuclear weapons exist and have been used, characters in all of these texts must assert their own ideas about survival rather than accepting the narratives of dominance.

After his father's death the boy must decide whether he will live by his father's values or his actions. After staying with his father's body for three days, he leaves and goes back to the road. When he sees someone coming he at first turns to hide, but then decides to stay and confront the stranger (already a deviation from the way his father acted, as they avoided contact with strangers if at all possible when he was alive). The man that approaches says the boy should come with him. Not knowing whether to trust him or not, the boy asks, "How do I know you're one of the good guys? You don't. You'll have to take a shot."[61] The man's reply highlights the uncertainty that the boy has faced and continues to face: not knowing whether he and his father were truly the "good guys" of the stories. The boy reveals his choice about how he will live through his decision to go with the man, and the narrative leaves readers to contemplate how these issues unfold in their own worlds. This text, with the final image of a little boy making his way in a harsh world, bestows readers with the sense that all three of these texts attempt to relate: that the stories we tell about nuclear issues reveal the relationship society has with their own past, and the decision, to accept the stories presented by dominant forces or create our own, reveals the way in which we hope to envision the future. The importance of stories and the issue of survival are themes that permeate all three texts, and the way in which *The Road* addresses these issues, by leaving the boy's future open, likewise leaves readers with a sense of hope despite the darkness of the text.

Conclusion

These texts reveal that although the nuclear referent's meaning has become more diffuse since the height of Cold War anxiety, it is still very much in use in contemporary society. Each narrative in this study incorporates concerns about the values we hope to live by through the way in which nuclear disasters lurk at the edges of these texts, and they show that much work remains to be done on how we deal with nuclear weapons in our world. The post-Cold War focus upon the aftermath of nuclear events highlights the role that nuclear weapons occupy in the cultural consciousness today, and the further concerns of these authors (government accountability, environmental threats, distrust of humanity) suggest areas that Nuclear Criticism would do well to consider in its continued role as a critical method for interpreting the world in which we live.

The nuclear threat can no longer be controlled by rhetoric about deterrence and Mutually Assured Destruction (M.A.D.), and this shift provides the opportunity to engage more truthfully with the issues at hand. As more countries hope to gain the power of the atom bomb, literature has much to teach us—if we are willing to listen. These three texts challenge previous models of nuclear literature through their emphasis on the importance of narrative and how the stories we tell about our nuclear past and possible nuclear futures reveal how we as a society deal with the use of nuclear weapons. Characters survive through reshaping the stories they have been told and through the development of the characters' sense of truth regarding the nuclear situations that structure these texts. They present a new model for thinking about the nuclear referent as we move forward in this nuclear age.

Notes

[1] Jacques Derrida, "No Apocalypse, Not Now (Full Speed Ahead, Seven Missiles, Seven Missives)," *diacritics* 14, no. 2 (Summer 1984), 23.
[2] Daniel Cordle, *States of Suspense: The nuclear age, postmodernism and United States fiction and prose* (Manchester and New York: Manchester University Press, 2008), 24.
[3] Ibid., 25.
[4] Editor's note: For further discussion of *Refuge*, see also in this volume Daniel Cordle's "Legacy Waste: Nuclear Culture After the Cold War."
[5] Derrida, "No Apocalypse, Not Now," 27-28.
[6] Cordle, *States of Suspense,* 26.
[7] Derrida, "No Apocalypse, Not Now," 30.
[8] Gerald Vizenor, *Hiroshima Bugi: Atomu 57* (Lincoln and London: University of Nebraska Press, 2003), 2.
[9] Atomic-bomb literature, as opposed to nuclear literature, is a genre which arose out of the need to comprehend the bombings of Hiroshima and Nagasaki. It has been predominantly composed by Japanese writers and for a Japanese audience, as the experience of an actual nuclear attack remains one that is unique to Japan. It is distinguished from nuclear literature because it is distinctly grounded in the event and aftermath of the atomic bombings of Hiroshima and Nagasaki, whereas nuclear literature deals with what Derrida calls "the massive 'reality' of nuclear weaponry and of the terrifying forces of destruction that are being stockpiled and capitalized everywhere," in Derrida, "No Apocalypse, Not Now," 23. This reality of the nuclear age is distinguished from the fiction of nuclear war, and this is the cause of the distinction between atomic-bomb and nuclear literature, for the reality of the nuclear age in Japan includes the reality, not fiction, of atomic attack. John Treat's *Writing Ground Zero: Japanese Literature and the Atomic Bomb*, published in 1995, provides a comprehensive study of atomic-bomb literature written by *hibakusha*, survivors of the atomic bombings, as well as writers of the post-war generation who continue to struggle with the after-effects of these atomic atrocities. In his critical study of atomic-bomb literature, Treat looks at how authors deal with events which "eclipsed comprehension [and] handicapped meaning" and the ensuing questions these events raise: "Can the causes—the reasons—be traced? Understanding achieved? Value restored? If so, to what end?" In John Whittier Treat, *Writing Ground Zero: Japanese Literature and the Atomic Bomb* (Chicago: University of Chicago Press, 1995), 3.

[10] Ineko Sata's "The Colorless Paintings," Mitsuharu Inoue's "The House of Hands," and Yoko Ota's "Fireflies," all published in Oe Kenzaburo's collection of atomic bomb literature *Fire From the Ashes: Short stories about Hiroshima and Nagasaki* (London: Readers International, 1985), are examples of this tradition.
[11] Vizenor, *Hiroshima Bugi*, 1.
[12] Ryunosuke Akutagawa's fiction deals with the ambiguities of human nature and what people are willing to do in order to survive. Akira Kurosawa's film derives from Akutagawa's "In a Grove." The film provides widely different accounts of a rape and murder, told by a woodcutter and a priest as they wait out a rainstorm. They reflect back upon the testimonies of a samurai, the samurai's wife, and a bandit. The stories are contradictory and can be seen to be motivated by personal gain. At the end, the "truth" is never made clear, and the film points to the subjectivity of truth itself.
[13] Vizenor, *Hiroshima Bugi*, 8.
[14] Ibid., 7.
[15] Ibid., 9.
[16] Ibid., 10.
[17] Ibid., 9.
[18] Ibid.
[19] Oe Kenzaburo, "Introduction: Toward the Unknowable Future," *Fire From the Ashes: Short stories about Hiroshima and Nagasaki* (London: Readers International, 1985), 15.
[20] Vizenor, *Hiroshima Bugi*, 81.
[21] Ibid.
[22] Ibid., 83.
[23] Ibid., 85.
[24] Ota Yoko was an author and *hibakusha* that survived the bombing of Hiroshima. She wrote *City of Corpses* on scraps of paper in the months following the bombing, as she felt an overwhelming need to tell her story about this experience. However, the authorities in occupied Japan censored the manuscript, which did not appear until 1948.
[25] Vizenor, *Hiroshima Bugi*, 88.
[26] Ibid., 36.
[27] Ibid., 22.
[28] Ibid., 205.
[29] Ibid., 206.
[30] Ibid., 18.
[31] Ibid., 20.
[32] Ibid.
[33] Terry Tempest Williams, *Refuge: An Unnatural History of Family and Place* (New York: Vintage Books, 1991), 3.
[34] Ibid., 23.
[35] Ibid., 39.
[36] Ibid., 43.
[37] Ibid.

[38] Rebecca Solnit, *Savage Dreams: A Journey into the Hidden Wars of the American West* (San Francisco: Sierra Club Books, 1994), 5.
[39] Ibid., 7.
[40] Ibid., 5.
[41] Williams, *Refuge,* 134.
[42] Ibid.
[43] Ibid.
[44] Ibid., 161.
[45] Vizenor, *Hiroshima Bugi*, 9, 36.
[46] Williams, *Refuge*, 137.
[47] Ibid., 4.
[48] Ibid., 283.
[49] Ibid.
[50] Ibid., 284.
[51] Ibid., 286.
[52] Ibid.
[53] For more on the prevailing imagery of nuclear winter, see also in this volume William Knoblauch's "The Pixilated Apocalypse: Video Games and Nuclear Fears, 1980 – 2012."
[54] Cormac McCarthy, *The Road* (New York: Vintage Books, 2007), 12.
[55] Ibid., 52.
[56] Ibid., 55.
[57] Ibid., 57.
[58] Ibid., 54.
[59] Ibid., 260.
[60] Ibid., 268.
[61] Ibid., 283.

Chapter Eight

Allegories of Hiroshima: Toward a Rhetoric of Nuclear Modernism

Mark Pedretti

Before and After Hiroshima

Like the blast wave that destroyed the city, the aftershocks radiating out from Hiroshima on August 6, 1945, crashed with amazing rapidity upon North American shores. No sooner had President Harry Truman announced the existence of the atomic bomb than the name "Hiroshima" came to signify an anxiety about the possibility of global nuclear destruction. At a time when the existing number of nuclear weapons could be counted on one hand, public discourse in the United States turned quickly to imagining future "Hiroshimas" and their inevitable conclusion in total nuclear conflagration. The atomic bomb had irreversibly altered not only the nature of modern warfare, but, according to Anne O'Hare McCormick, writing in the *New York Times* only days after the bombing, it had also triggered "an explosion in men's minds as shattering as the obliteration of Hiroshima,"[1] a sentiment that historian Paul Boyer reminds us "was echoed by literally scores of observers in these earliest moments of the atomic age."[2] Thoughts quickly turned from the "city of ruins and dead"[3] that Hiroshima—and, three days later, Nagasaki—had become, to "the horrible prospect of utter annihilation opened by the atomic bomb."[4] Only one day after the bombing, the *St. Louis Post-Dispatch* was warning that the Manhattan Project may have "signed the mammalian world's death warrant and deeded an earth in ruins to the ants."[5]

Long before the absurdity of the Cuban missile crisis or the farce of Ronald Reagan's faux-epic struggle with the "Evil Empire," these earliest moments of the atomic age could still take seriously the experience of nuclear anxiety; however impossible total nuclear war was in the days (and months and years) following Hiroshima, the event of the bombing nonetheless produced in American public consciousness a sincere and newfound "sense of an ending" predicated on a drastically foreshortened future.[6] It is this sense of futurelessness, most of the major newspaper commentators agreed, that made the advent of nuclear technology a watershed moment in human history; as Barnet Nover wrote in the *Washington Post*: "On the morning of August 6, 1945, the world as we have long known it disappeared in the dust and flame that rose in a gigantic cloud over Hiroshima, Japan... The particular bomb that fell on Hiroshima literally transformed the world."[7]

This trope of a "new world" or a "new age" dawning was repeated in hundreds of variations, to the extent that it joined the laundry list of clichés which constituted, according to Spencer Weart, "the whole tangle of fantastic nuclear imagery... [which] was a significant part of the mental equipment of everyone within reach of a radio."[8] Nevertheless, we seem to have taken those claims to novelty seriously, as the end of World War II has conventionally come to designate a turning point in Western geopolitical, social, and economic configurations. The study of American literature is apparently no exception, as the year 1945 conveniently punctuates the periodization of the twentieth century, clearly delineating the modern from the postmodern (whether understood as aesthetic practice, ideology, or *episteme*). Indeed, the prevailing heuristic of literary periodization has been to align the nuclear age with American postmodernism. Sarah Henstra has recently summarized this critical consensus:

> In fact, many theorists of postmodernism have turned to nuclearism and Cold War fear to account for the peculiarities of their subject: fragmentation, desperation, fantasy, relativism, futility, the play of language in the face of an absurd or absent "reality." While it is undoubtedly too reductive to *equate* the postmodern with the nuclear, the development and proliferation of nuclear weapons does make the postmodern era something fundamentally distinct and unique than merely the aftermath of, or an hiatus from, modernism (as the suffix "post" might suggest).[9]

Henstra's equation of postmodernism and nuclearism is confirmed by the mainstays of Nuclear Criticism,[10] including a 1984 conference at Cornell sponsored by *diacritics*, Alan Nadel's seminal *Containment Culture: American Narratives, Postmodernism, and the Atomic Age* (1995), Margot Henriksen's *Dr. Strangelove's America: Society and Culture in the Atomic Age* (1997), as well as more recent incarnations like Daniel Cordle's *States of Suspense: The Nuclear Age, Postmodernism and United States Fiction and Prose* (2008, and whose work is also included in this volume) and Daniel Grausam's *On Endings: American Postmodern Fiction and the Cold War* (2011), among many, many others.[11] This litany of texts, I would argue, shares certain axiomatic assumptions about twentieth-century literary periodization; not only is the historical coincidence of the Cold War and postmodernism taken as a *prima facie* warrant for positing a correlative or causal relation between them, but, following Henstra, including the atomic bomb in the calculus of the postmodern characterizes the period in relation to a preceding modernism in terms of a qualitative break or rupture. The technological novelty of the bomb is thus superimposed onto the literary-historical epoch that produced it and which it defines, such that postmodernism too becomes predicated on the advent of the new. The claims of a "new world" made in the days after Hiroshima become the signal feature of the epoch which the event itself engenders, and, despite insistences to the contrary, the lines of causality start to run solely in one direction. As Tony Jackson writes, "It is in this way that from early on in the fifties a certain atmosphere emerged as a result of the creation of weapons that could bring about an absolute end. I do not claim that deterrence thinking straightforwardly caused poststructuralist (or more broadly, postmodern) conceptuality to emerge, but I do hold that the Cold War atmosphere was instrumental in enabling such conceptuality to appear and thrive as it did."[12]

But is this periodization schema the only, or even the most productive, one for considering early Cold War literary production? To be sure, the postmodern/post-nuclear equation (to borrow Rob Wilson's formulation)[13] has much to recommend it, and it undoubtedly functions best in accounting for the canonical "high" postmodern fiction of the late sixties and onward (Pynchon, Barth, DeLillo, et al.). Yet, as with all periodizing concepts, it obscures as much as it reveals, foregrounding a particular set of questions and concerns at the expense of others. If, as Fredric Jameson has contended, the individual work exists in a dialectical relationship to the periodizing concept under which it is treated, then the text will "show up" according to the set of generic presuppositions that we bring to it, even as the text's resistances to that concept might point towards alternative possible periodizations.[14] And while no periodizing concept is wholly determinative for the text to which it is applied (it can only function, as Brian McHale has argued, as "the dominant" organizing and interpretive principle),[15] the limitations of the postmodern/post-nuclear alignment become evident when trying to account for the literature of roughly the first two decades of the atomic age—a group of authors and texts which, taken together, do not easily conform to the rule-of-thumb litmus tests of postmodernism. How, for instance, would we square postwar—indeed, Cold War—novels such as Ralph Ellison's *Invisible Man* (1952) or Sylvia Plath's *The Bell Jar* (1963), both deeply concerned with the possibility of knowing the world, with something like McHale's epistemological/ontological distinction as a way of separating the modern from the postmodern?[16] It strikes me as remarkably difficult to make a case for novels such as these (not to mention the Beats, confessional poets like Robert Lowell, or the later works of "high" modernists like T.S. Eliot or Faulkner) in terms of an ontological emphasis that would signify a postmodern aesthetic. And yet, if we are to align a synchronic periodizing concept (style) with a diachronic one (historical event), as the nuclear critics have done, then we are left with little alternative but to force impossible stylistic constraints onto texts by virtue of their historical location.

In this chapter, I seek to break the aesthetic-historical synchronicity that has guided Cold War literary studies by arguing that the decades immediately following World War II might be better treated under a periodizing rubric of *late modernism*, what Jameson has called a "transitional concept" capable of accounting for the both anomalous and asynchronous nature of cultural production in relation to historical formations, and of making "some place... [for] the last survivors of a properly modernist view of art and the world... who had the misfortune to span two eras and the luck to find a time capsule of isolation or exile in which to spin out unseasonable forms."[17] Thinking about the advent of the atomic age in terms of late modernism resolves the dilemmas of periodization discussed above (and to which I shall return), while at the same time suggesting a different set of interrogative coordinates for describing this period. Instead of treating the immediate postwar period in terms of postmodern "fragmentation, desperation, fantasy, relativism, futility, the play of language in the face of an absurd or absent 'reality,'"[18] we might instead consider it as the persistence of a modernist aesthetic *in spite of* changing historical circumstances; that is, the postmodern turn to irony and pastiche seems to me a second-order effect, a cutting of ideological and aesthetic losses, only *after* a modernist project has been thoroughly abandoned in the recognition of its own futility or obsolescence. If we take modernism, broadly construed, as the desire for an open and transformable future, and if the futurelessness engendered by the atomic bomb radically thwarts that desire, we nonetheless would not conclude that such a desire disappears overnight; late modernism would be the name for that interregnum of frustrated desire, a utopianism deprived of the prospect of utopia.[19] Instead of asking, with Jean Baudrillard, "What are you doing after the orgy?,"[20] the late modernist perspective would ask, with James Agee's article for *Time* magazine in the weeks following Hiroshima, "In an instant, without warning, the present had become the unthinkable future. Was there hope in that future, and if so, where did hope lie?"[21] Agee's question, in all of its modernist valences, suggests that it might be more productive here to speak of a certain strain of what I am calling nuclear modernism, a particular subspecies of late modernism in general which has heretofore gone unaccounted.

Arguing for nuclear modernism in this way represents more than a splitting of hairs over periodization. As I will suggest below, the conceptual coordinates associated with late modernism are markedly different from those of a full-blown postmodernism, and its interface with the nuclear era allows us to ask, perhaps for the first time, a different set of questions about the form and movement of historical time and the possibilities of historical transformation. The postmodern Nuclear Criticism of the 1980s presented an aporetics of historical time; the proper tense of the nuclear age, according to Richard Klein,[22] would be a "new, nonnarrative future tense" that takes the form of a "negative future anterior," what *will not have been*.[23] The structure of such a "futureless future," however, remains for Klein just beyond the horizon of cognition: "*If* [the nuclear age] *could* grasp the structure and implications of this New Future…"[24] While the remainder of Klein's essay somewhat gnomically attempts to articulate what such a tense might look like, he has already constructed an unsurpassable temporal aporia: if the nuclear age is "new" in that it requires a "new future tense," then the *event* of its emergence— its rupture with the past—either must be temporalized according to the old tenses which the event modifies, which means that it is not new after all; or it must be temporalized according to that "new future tense," which would then need to precede its own historical emergence. Such an aporia, I would contend, appears whenever one takes as axiomatic the idea that the atomic bomb represents a decisive historical break or rupture—or, indeed, that "break" and "rupture" are meaningful categories of historical analysis. Indeed, from this vantage we could understand the postmodern turn to historical fragmentation and textuality as a resignation in the face of this aporia, a throwing up of the hands in trying to work through the problem of historical transformation.

By contrast, if we start with the commonsense assumption that modernism had not "unexpectedly vanished without a trace" on August 7th, 1945, then we must admit to a transitional concept like nuclear modernism, which nonetheless allows us to consider how a modernist historical imaginary confronts the apparently real prospect of imminent nuclear ending.[25] Such an imaginary would be committed to "what was likely to come of such changes and their general tendency; [it] thought about the thing itself, substantively, in Utopian or essential fashion," rather than the "fabulously textual" futurity of postmodern Nuclear Criticism.[26] As Agee's question suggests, and as I hope to demonstrate, late modernist authors certainly took the prospects of their own future very seriously, and attempted to repurpose a modernist vocabulary of historical change to describe the nuclear threat; but, importantly, they did not abandon the future as such. As that utopian imagination butts up against the limits of its own historical possibility, it also comes most clearly into view. If "the question of Utopia would seem to be a crucial test of what is left of our capacity to imagine change at all," then a Nuclear Criticism that could retrieve or rearticulate such a possibility seems a worthy endeavor.[27]

Besides trying to carve out historical space for such a nuclear modernism, my second aim here is to suggest how its rhetoric might begin to be organized. I contend that the gesture of the journalists and pundits who first reported on Hiroshima offers some guidance. When confronted with the destruction of the city, they immediately projected into the future a bigger, more destructive bomb, effectively erasing the specificity of Hiroshima as a place or an event, turning its horror into a fear for self-preservation. This is what psychologist Robert Lifton contends Americans usually do when confronted with the magnitude and guilt of the bombings: avoid, displace, and dissociate. Writes Lifton,

> One was supposed to be numbed to Hiroshima. It became politically correct (before the expression existed) in the deepest sense to remain numbed toward Hiroshima—politically suspect if one was troubled or inclined to make a fuss about it. In that way, as a people, we developed a *habit* of numbing toward Hiroshima, a sustained tendency toward, one way or another, avoiding feeling in connection with what happened there... One can say that Americans have generally possessed a vague, unfelt, half knowledge of Hiroshima—usually of a kind that interfered with a more general capacity for grasping nuclear-weapons truths. But that does not mean we are free of fear.[28]

However we might judge the activity of turning the victims of Hiroshima into symbols, we can understand this dissociative process as the *allegorization* of the event in the de Manian sense of a displacement or a deferral, "a distance in relation to its own origin, establish[ing] its language in the void of this temporal distance."[29] That is, Hiroshima as an event has no meaning in itself, only that with which it is allegorically invested, following Jameson, through "successive rewritings and overwritings which are generated as so many levels and so many supplementary interpretations."[30] The meanings given to the event (as retribution for Pearl Harbor, as harbinger of the end of the world, as dawn of a new age, etc.) constitute allegories which serve to cover over its original trauma—for both its victims and perpetrators.[31] And to the extent that this process of allegorization ignites such displacements, it also conditions what Walter Davis has referred to as the "nuclear unconscious" of both ourselves and the literary text, thus allowing us to read in texts that do not take the nuclear as an explicit referent the traces, like a radioactive half-life, of the atomic bomb. In this sense, allegory might be said to constitute the master trope of nuclear modernism, and thus offers a mode of textual interrogation that begins looking for the nuclear precisely in its avoidances and denials. Allegory would represent a particularly powerful instance of a distinctly late modernist relation to temporality—asynchronous and belated—that represents the distance between the historical present and the concepts of futurity by which it is understood.

I chart here two texts that I think represent opposite ends of a spectrum of such avoidance: John Hersey's quintessential nonfiction novel *Hiroshima* (1946), and a less well-known text by John Hawkes, *The Cannibal* (1949). These texts register the initial shock of Hiroshima, and attempt to reckon with it within a language of modernist convention. But they approach this task in very different ways, constructing a dialectic of form and content which may prove useful for the consideration of other late modernist texts as well. It would appear from Hersey's and Hawkes's texts that, when it comes to representing the magnitude, the horror, or the implications of Hiroshima, novelistic form and diegetic content exist in a mutually exclusive relation, where it seems impossible to adequately capture one without sacrificing the other. Hersey's *Hiroshima*, for instance, offers one of the first, and still most commonly known, accounts of the bombing, based on the testimony of survivors; but its flat, journalistic objectivity seems to fail to do justice to the deracination that those survivors experienced. Hawkes, on the other hand, pushes the limits of narrative form, as an extension of a modernist experimental tradition, and seems to capture something essential about the incipient sense of futurelessness generated by the bomb; but he is only able to do so by displacing the event of Hiroshima itself on to a story about Nazi Germany. In this sense, the effects of Lifton's "psychic numbing" can be seen as unavoidable in one way or another, leaving us to trace instead the displacements of particular texts and chart their rhetorical strategies. Such an approach might then begin to adumbrate a domain of nuclear modernism, and to see it as a crucial, distinct hinge moment in twentieth-century literary history.

Restoring History: John Hersey's *Hiroshima*

First published in *The New Yorker* in August 1946, John Hersey's journalistic account of the bombing was, for many Americans, their first and only cognizance of Hiroshima; it was quickly reprinted in various national and regional newspapers, distributed freely to the Book-of-the-Month Club, and read aloud on national radio over four successive evenings.[32] According to Michael Yavenditti, Hersey's graphic depictions of the effects of the explosion, coupled with his approach of recounting the stories of six survivors, worked to humanize the Japanese citizenry for an American public desensitized by five years of brutal war. As such, argues Yavenditti, *Hiroshima* "prompted some Americans to rethink their previous approval of the atomic bombings, while it intensified the anger of those who had initially condemned the bomb's use," even if, in the long run, Hersey's piece "aroused many readers but inspired few of them" to actively protest emerging U.S. nuclear policy.[33] Spencer Weart observes that the lasting effect of Hersey's personalized perspective was to make Americans realize that the bomb had been used "against people like ourselves," and thus could just as easily be used at home.[34] Even today *Hiroshima* remains a staple of high school and college curricula, serving variously as an exemplar of the New Journalism, the nonfiction novel, or a study in historical trauma.

Despite its widespread dissemination, early critics of the novel saw within it a more general problem of representing the horror and magnitude of the bombing at all. Mary McCarthy, for instance, argues that the journalistic objectivity of Hersey's prose undermined the radical novelty of the bomb, the ways in which this event could not be assimilated to previous forms of human experience. "What [*Hiroshima*] did," she writes to Dwight Macdonald in *Politics* magazine, "was to minimize the atom bomb by treating it as though it belonged to the familiar order of catastrophes—fires, floods, earthquakes—which we have always had with us and which offer the journalist... an unparalleled wealth of human interest stories, examples of the marvelous, and true-life narratives of incredible escapes."[35] According to McCarthy, to attempt to represent Hiroshima is to expose it to conventionalized (and conventionalizing) discourse, thus rendering the event "familiar and safe, and so, in the final sense, boring."[36] McCarthy would rather preserve the "kind of hole in human history" that Hiroshima represents, which seems to entail an abiding silence around it.[37] In a later consideration, Kingsley Widmer judges Hersey's project along the same lines but even more harshly:

> Given the scale of the events and the refusal of individual focus, the fictional verisimilitude (right down to brand names and a nearly surreal over-exactness on odd details) must falsify the reality. The author's pose of merely recording and reporting allows him to obscure personal evaluation, attenuate intellectual reflection, abstain from moral argument and abnegate poetic mediation. Such decorous and unthoughtful fictionalized reportage illustrates not only the limitations of commercial authorship but of the popular audiences and our standard exploitative publication which combines over-indulgence with undernourishment. The documentary-novel is itself a technological exploit; it expertly and vividly "communicates" but at the price of a considerable reduction in human response and relevance... Such is John Hersey's *Hiroshima*.[38]

For Widmer, an event such as Hiroshima poses the insistent question, "Is the horror and guilt so great that no thought and imagination can be made relevant?"[39] The event demands a literary response, but any attempt to represent it "can only provide a paltry, and probably falsified, reality."[40] Hersey, on the contrary, suggests elsewhere that he considered himself up to the task of representing the event by engaging in "an honest effort to comprehend the forces of contemporary history" through the "clarifying agent" of prose[41]—and he made no secret that the novel's emplotment followed the modernist influence of, among others, Thornton Wilder's *The Bridge of San Luis Rey* (1927).[42] What appears to Hersey as a problem capable of being solved we might understand as a fundamental dilemma that organizes the representation of Hiroshima at large: the event at once requires representation, but dwarfs any and all attempts to do so.

Such an aesthetics-of-failure, of course, sounds a great deal like the premise for a work of postmodern irony and self-reflection; this is the kind of reading of *Hiroshima* which Nadel has advanced, marking one of the definitive instances of the postmodern/post-nuclear alignment I challenged earlier. Thus, revisiting Nadel's interpretation offers the opportunity to demonstrate the limits of that periodization schema more generally. Like the rest of *Containment Culture*, Nadel's reading of Hersey aims to demonstrate how the structures of rhetorical and ideological containment inevitably fissure before the unknowability of nuclear destruction, thus laying the foundations for a postmodernism that celebrates "a self-referential awareness of historicity and artificiality and a cognizance of the fissure between 'history' and 'event.'"[43] At different points Nadel sees *Hiroshima* as either merely symptomatic of "the *formal* failure of containment," or self-consciously aware of its textuality and textually-produced authority, in a "subtle anticipation of the principles of postmodernism."[44] This equivocation seems important, because it marks the boundary between viewing *Hiroshima* as late modernist and early postmodernist. Initially, Nadel demonstrates how Hersey's text promises to construct a totalizing narrative about the "truth" of the bombing, only to have the event continually exceed or elude such totality; if narrative requires authority, organization, and causation, then the bombing undermined traditional sources of authority and knowledge, foregrounded random chance, and "destroyed the ability to apply accepted notions of causality, to make traditional distinctions, or to classify events historically."[45] Nadel points to no less than five mediating frames between Hersey's narratorial voice and the event itself (memory, language and cultural barriers, etc.), all of which are omitted—undoubtedly along with untold amounts of primary source material—for the sake of narrative economy and form. But then Nadel advances a stronger claim, that Hersey is *aware* of the artificiality implied by such omissions, thematizing "the arbitrariness of the frame, the failure of metanarratives, and the instability of writing."[46] The remainder of his reading—amply supported, to be sure—proceeds to illustrate the "important assumptions of postmodernism" at work as Hersey's text foregrounds its "self-consciousness, its cognizance of itself *as document*" and the artificiality of its authority, for which actual historical testimony acts "only as a supplement."[47] For Nadel, *Hiroshima* becomes the incipient moment of an emergent postmodern aesthetic and epistemology, where the first few cracks in a containing metanarrative appear.

This interpretation has much to recommend it, but, as I suggested, it also demonstrates what is lost—both specifically and generally—in locating such an early postwar text on the side of the postmodern (assuming we follow Nadel's stronger claim of self-conscious textuality). First, as Hersey's comments above imply, he does not seem to share Nadel's estimation of the reflexive foregrounding of artifice, irony, and representational failure; on the contrary, Hersey imagines his writing as attempting to see "if the horror could be presented as directly as possible."[48] Whether he succeeds or not is another question; but it seems meaningful to distinguish between *trying* to represent the event and coming up short, and something like Samuel Beckett's "Try again. Fail again. Fail better," in which failure itself becomes the goal of literary representation.[49] (On a related note, Nadel seems to misunderstand Hersey's generic innovation of the nonfiction novel as working "in the name of truth," since Hersey saw himself as adding the devices of fiction to the genre of journalism, not the other way around: "I had experimented with the devices of fiction in doing journalism, in the hopes that my mediation would, ideally, disappear.")[50] While this sort of authorial pronouncement is by no means definitive, it does provide a useful indication of how the writer imagines his own project. And in this case it would be worth observing that Hersey, along with other writers who flourished in the postwar years, saw themselves as the inheritors of a distinctly modernist tradition, permuting and adapting the formal innovations of the interwar generation to meet an evolving historical reality.

Second, the force of Nadel's reading is to make *Hiroshima*, as it were, pre-postmodern: the novel is "a subtle anticipation" for a "legitimation crisis between scientific knowledge and narrative knowledge that has *not yet* become visible as a crisis."[51] In this estimation, its literary value derives from what it foretells, *not from what it is*; the text becomes important only insofar as it refers to a subsequent literary period to which it nonetheless remains marginal. Such instrumentalization tends to neglect the particularity of late modernism in general, leaving it, as Tyrus Miller has written, "little more than a peripheral issue, a bit of detail work on the capacious but drafty house of fiction built by Modernism, Postmodernism, and Co."[52] Similarly, Andrew Hoberek has discussed this kind of neglect bestowed upon postwar literature more broadly: "On the whole scholars treated the decade [the fifties] like one of those big Midwestern states you have to go through to get somewhere else: a literary and cultural Kansas taking up space between the triumphs of modernism and the postmodern revival."[53] Nadel, for instance, considers Hersey's formal innovation—the nonfiction novel—only as a straw man to be readily knocked down in the (postmodern) terms that it also engenders. Admittedly, narrating the late modernist moment lacks something of the energy and exuberance that inheres in even garden-variety accounts of either "high" modernism or postmodernism, but that does not mean that there is a not a story there to be told, on its own terms, and which has the potential to reposition the literary periods that come before and after it.

This reassessment of Nadel's reading maps some of the major coordinates for periodizing late modernism as a distinct moment of transformation in twentieth-century literary history. We would have to start with the assumption that, whatever retrospective hermeneutics we might bring to the table in light of a later generation of postmodernists, the writers of the late forties and fifties understood themselves to be contiguous with an established lineage of American modernism, elaborating on its lineaments of formal innovation rather than breaking with them. But then we would have to observe how, in light of an emergent experience of futurelessness produced by the atomic bomb, the salient features of a modernist ideology of novelty begin to lose their explanatory purchase. As such, the late modernist moment might be seen as refashioning the techniques of modernism to suit that emerging historical reality. And that effort may, more often than not, result in a certain kind of failure or inadequacy, but without self-consciously valorizing or fetishizing it as the end of representation.

From this vantage, John Hersey's *Hiroshima* looks less like an anticipation of postmodern indeterminacy than a vigorous effort to assert formal coherence in the face of the deracinating experience of the bombing. As the quotations from McCarthy and Widmer above suggest, Hersey's text fits uneasily with its represented object; the magnitude, the horror, and the novelty of the bombing continually exceed the flat, conventional realism of his style. I would follow Nadel this far, as *Hiroshima* repeatedly manifests its own inadequacy for capturing the sense of sudden and complete world-ending commonly reported by the survivors,[54] as well as the prevailing sense of futurelessness of the atomic age more generally. But this manifests itself in the text not as a dim self-awareness but as a neurotic symptom, a compulsion to *deny* that reality through narrative substitution. If the event of the bombing introduces an altered model of temporality deprived of a future, then *Hiroshima* attempts to *restore* linear chronology, to reestablish historical time according to the ideology of modern progress, as homogeneous, unidirectional, and unending. In this sense, Hersey's text is allegorical in Jameson's sense of an "overwriting," as the temporality of narrative representation displaces the temporality of the event, containing its alterity within the bounds of an everyday, linear time.[55]

Hiroshima begins by invoking a temporal frame: "At exactly fifteen minutes past eight in the morning on August 6, 1945, Japanese time, at the moment when the atomic bomb flashed above Hiroshima."[56] There seem to be two concomitant and contradictory gestures here: first, the articulation of a distinctly "Japanese" time, which promises in its Oriental exoticism something other than the linear clock-time of the West; but then, as soon as that alternate temporality is suggested, it is subsumed back into an "exact" chronology. If there is a "Japanese time" in the text, it is mentioned only to be instantly wiped away, as if the bombing itself produced the temporal order of American military time. In fact, Hersey's opening repeats, turn for turn, the language of the U.S. Strategic Bombing Survey's report to President Truman: "A single atomic bomb, the first weapon of its type ever used against a target, exploded over the city of Hiroshima at 0815 on the morning of August 6, 1945."[57] It is possible that

Hersey's invocation is parodic,[58] but I would contend that the perspective *Hiroshima* offers is that of the bomber *Enola Gay*: an aerial, cartographic perspective that also implies the selection of military targets. Early in the text, we are given such a bird's-eye description of the city, immediately after "an American weather plane [actually a targeting surveyor] came over:"

> Hiroshima was a fan-shaped city, lying mostly on the six islands formed by the seven estuarial rivers that branch out from the Ota River; its main commercial and residential districts, covering about four square miles in the center of the city, contained three-quarters of its population, which had been reduced by several evacuation programs from a wartime peak of 380,000 to about 245,000. Factories and other residential districts, or suburbs, lay compactly around the edges of the city. To the south were the docks, an airport, and the island-studded Inland Sea. A rim of mountains runs around the other three sides of the delta.[59]

The text positions us above the city, in a neutral, universal timeframe, leveling off potentially disparate temporalities under Walter Benjamin's "homogeneous, empty time."[60] This encyclopedic description becomes essential for understanding the subsequent movements of the characters; as the narrative voice for the most part inhabits the perspectives of its subjects, following their movements around the city, with this geography we are able to reconstruct their locations quite easily. So when Father Kleinsorge, one of Hersey's "characters," walks with his fellow Jesuit missionaries "along the Ota above the city," crosses "Misasa bridge," and encounters a group of soldiers "in the center of the town," we have little difficulty charting this journey.[61] Likewise, Mr. Tanimoto's ferrying of the wounded across the river to Asano Park fits into what we already know about Hiroshima's tributary system.[62] This spatialization suggests another central modernist trope which Hersey is adopting: like the games we can play with texts like *Mrs. Dalloway* or *Ulysses*, mapping the routes of Clarissa Dalloway or Leopold Bloom around their respective metropolises, Hersey asks us to reconstruct the city based upon his characters' subjective encounters with the urban environment.

Throughout the rest of his narrative, Hersey will continue to insist rigorously upon this linear chronology, marking most of the chapters, sections, and transitions with these temporal designators: "Immediately after the explosion," "right after the explosion," "early in the evening of the day the bomb exploded," "by nightfall," "early that day, August 7th," "before dawn on August 8th," "on August 9th," "on August 18th, twelve days after the bomb burst," and so on.[63] We are told early in the text that

Hiroshima marks "the first moment of the atomic age,"[64] and that Hersey's text promises to explain the novelty of the event and its attendant technology—the editors of *The New Yorker* justified their decision to devote an entire issue to the piece on the similar grounds that "few of us have yet comprehended the all but incredible destructive power of this weapon"[65]—but it does so by quickly reducing it to the familiar temporal order of the clock and the calendar.

In 1985, Hersey added an epilogue to *Hiroshima* to commemorate the fortieth anniversary of its publication. Entitled "The Aftermath," this new chapter elongates the linear temporality established in the original narrative by documenting in brief the intervening forty years of nuclear history. The title of the section by itself assures us that time has continued its steady march. For there to be an aftermath there must be an "after," suggesting that the initial apprehensions of futurelessness have been safely subdued. And, just as the original text was punctuated with temporal designators marking the passing days after the bombing, so is the chronicling of the six survivors' subsequent lives marked, but on a larger scale. With the same insistence, Hersey's paragraphs begin: "In 1951," "in 1966," "in later years," "in 1972," "by 1977," "New Year's Eve, 1963," and so on.[66] Marking the stretched time of a life in this way, the text emphasizes the survivors' endurance as the initial terror of the event recedes further and further into the past. For the most part, the time scale of this annualized chronicle remains steadfastly personal; world-historical events are only relevant insofar as they bear on the biographies of the survivors. But then, in the final sketch of Kiyoshi Tanimoto, the narrative is interrupted by italicized reports of significant developments in the global history of nuclear weapons: the Crossroads tests of 1946, the first Soviet bomb test in 1949, up through India's announced proliferation in 1974. At this point, world history threatens to take the place of biography. Unlike the necessary finitude of an individual human life, the steady progression of a universal time frame measured in years assures the reader that time goes on indefinitely, returning us to, or restoring for us, the endless linearity of modernity.

With this conflation of the personal and the global in mind, and understanding how the text's persistent assertions of normalcy speak to its latent anxieties, the final paragraph of the epilogue seems an equivocal gesture:

> Kiyoshi Tanimoto was over seventy now. The average age of all hibakusha was sixty-two. The surviving hibakusha had been polled by *Chukogu Shimbun* in 1984, and 54.3 per cent of them said they thought that nuclear weapons would be used again. Tanimoto read in the papers that the United

> States and the Soviet Union were steadily climbing the steep steps of deterrence. He and [his wife] Chisa both drew health-maintenance allowances as hibakusha, and he had a modest pension from the United Church of Japan. He lived in a snug little house with a radio and two television sets, a washing machine, an electric oven, and a refrigerator, and he had a compact Mazda automobile, manufactured in Hiroshima. He ate too much. He got up at six every morning and took an hour's walk with his small woolly dog, Chiko. He was slowing down a bit. His memory, like the world's, was getting spotty.[67]

Manifestly, Hersey intends for this final stroke to be foreboding, a reminder that the nuclear danger has not (in 1985) subsided but escalated, with the very real possibility of history repeating itself. And yet, as the text shifts focus between Tanimoto and the world, we can just as easily read this paragraph as sign of the vanquishing of nuclear temporality and the restoration of futurity. The adequation of personal and global in the last sentence allows Tanimoto to stand in for all *hibakusha* (in Japanese, explosion-affected people), if not for humanity in general; and in that regard, humanity promises to endure. Tanimoto has lived longer than he statistically should have, driving a car manufactured in a rebuilt and modernized Hiroshima. Forgetting here does not promise historical return or repetition, but rather absolution and erasure. If the "hole in history" which marks a before and after is dissipating on the foggy shores of historical memory, then Hersey's text, in its final move, sutures historical time back together into an unbroken continuity. The prospect of a radically different temporal order, predicated not on progression but imminent ending, has been symbolically neutralized within a narrative form that insists upon an abiding linearity.

Elision and Anachrony: John Hawkes's *The Cannibal*

Hiroshima represents one end of a spectrum, I am contending, for how American late modernism confronts the diminished futurity of the bomb. To be reductively schematic, Hersey gets the content right but the form wrong (at least on the level of the text's overt presentation); he represents the bombing directly, but through a narrative mode, heavily indebted to a canonical modernism, that does violence to the event. Nonetheless, that formal violence reveals a crucial operation of the text's "nuclear unconscious." At the other end of the spectrum, I would locate John Hawkes's *The Cannibal*, a novel published only three years after Hersey's. Like Hersey, Hawkes participates in a modernist tradition of formal experimentation, seeking to further push the limits of novelistic discourse; but unlike Hersey, Hawkes in *The Cannibal* allegorizes the nuclear condition in a narrative of occupied Germany, instead attempting to instantiate futurelessness at the level of form. Set on a dark night in April 1945, *The Cannibal* is narrated by a man named only Zizendorf who, with a pair of bumbling co-conspirators named Stumpfegle and Fegelein, are waiting by the side of the only road into the fictional town of *Spitzen-on-the-Dein* to assassinate the lone American overseer of this portion of Germany, a motorcyclist named Leevey, and thus incite a neo-Nazi revolution that will recapture Germany's lost martial past. Just before Zizendorf's plan is to unfold, the text jumps back to 1914, and an almost entirely unrelated narrative about the courting of a young woman named Stella Snow by her suitor Ernst, before returning in the novel's final movement to the scene of Zizendorf's ersatz revolution. It is this jumping back and forth, along with the deferral of its ostensibly central event, by which Hawkes attempts to construct a distinct temporality for the nuclear age.

Despite its setting, there is something apocalyptic about *The Cannibal* that conjures post-nuclear fantasies of total destruction. The town of *Spitzen-on-the-Dein*, also called "*das Grab*" or "The Grave," is, quite literally, a ghost town, vaporized by the effects of war and occupation.[68] Hawkes's extensive descriptions of setting pay particular attention to the smells of destruction,[69] and the ways in which, in a symptomatic allusion to Eliot, the town has become a waste land, unable to support life:

> All during the day the villagers had been burning out the pits of excrement, burning the fresh trenches of latrines where wads of wet newspapers were scattered, burning the dark round holes in the back stone huts where moisture traveled upwards and stained the privy seats, where pools of water became foul with waste that was as ugly as the aged squatter. These earthen pots were still breathing off their odor of burned flesh and hair and biddy, and this strange odor of gas and black cheese was wafted across the roads, over the fields, and collected on the damp leaves and in the bare night fog along the embankment of the *Autobahn*.[70]

What few human inhabitants remain are joined by the ghosts of English soldiers from a burnt-out tank on the road, "bodies that slowly appeared one by one from the black foliage" and "who left it at night and hung over the canal walls for drink."[71] Images of depopulation abound: "the undertaker had no more fluid for his corpses," and "the halls no longer smelled of roasting swine or boiling cabbages, no longer rang full with heavy laughter, but remained dark and cold, streaked with mud from the roomers' boots."[72] The land has also become sterile, full of "bleached plants," "endless grey fields," and "twisted stunted trees"; "the green of cabbages had turned to white."[73] On top of that, the four horses of Biblical apocalypse are roaming around,[74] and in the final section of the novel, simply entitled "Three," it does not require much to imagine, as Patrick O'Donnell has put it, "another, perhaps final and cataclysmic outbreak of chaos and war."[75] In his introduction to the novel, Hawkes's mentor Albert Guerard compiles these images of death, mutation, and decay to suggest that they amount to the generalized fantasy of annihilation hanging over the early postwar West:

> This is a Germany of men with claws for hands, of women with reddened flesh, of children with braces to support their stumps or heads. It is a world without food, without hope, without energy… reduced for its pleasures to impotent mechanical rutting bereft of all desire. I think it can be understood that this is more than post-war Germany, whatever the author intended; that this is, to some degree, our modern world.[76]

That *The Cannibal* eschews ostensive reference to the nuclear age and yet so obviously invokes it allegorically suggests that its futureless temporality must instead appear on the level of narrative form. If the imagination of total nuclear destruction refuses what Peter Brooks has termed the hermeneutic "anticipation of retrospection" of all narrative—looking forward to looking back once we have finished the text[77]—by positing an end where no retrospection is possible, then we would expect such a formal engagement to rely upon other narrative strategies to disrupt our usual modes of readerly anticipation. *The Cannibal* does this through extended use of analepsis and prolepsis, which, as Gérard Genette reminds us, are rhetorical figures of anachrony that can both produce and heal "temporal rupture."[78] For Hawkes, these devices work in terms of large-scale textual structuration all the way down to the level of the sentence. While they pervade nearly every plot thread and scene of the novel, for the sake of brevity I will focus on one central instance: the death of Stella's Snow's mother when an airplane falls out of the sky and crashes into her. This event is narrated in what Genette would call the "zero degree" of correspondence between story-time and discourse-time[79] in the "1914" section of the novel, about one-third of the way through the text, but we have learned about it as early as page twelve, in the "1945" section, when Stella recalls "when her mother fell before her in the street from marketing, a piece of metal jutting from the bosom, while the airplane crashed."[80] This analepsis is complimented by a prolepsis early in the "1914" section, before the plane crash, when Stella has a vision of "the future fluttering wings of the solitary British plane leaving its token pellet in the market place, her mother's body rolling around it like a stone stained forever."[81] Both of these anachronies come from Stella's consciousness, but they also infect the narrative discourse, penetrating into ostensibly unrelated story-lines. For instance, as a mad "Duke" chases a boy around the town in "1945," "about him the wind began to scream as through the slots of airplane wings."[82] After the event, an unusual amount of narrative attention is paid to wings of all kinds, as when, at her mother's funeral, Stella "heard the perpetual scratching feet of insects who walked over the coffin lid with their blue wings" while "it was so hot that the birds buried their heads in the shade under their wings."[83] As these analepses and prolepses echo through the text, when the event is actually narrated in the "zero-degree," it lacks any sense of narrative surprise or anticipation, appearing only as a bland determinism, a death foretold without much interest. When the scene unfolds, we might notice certain details that would otherwise appear trivial—"a speck appeared in the sky... A few loiterers got out of their way"—but we already know how it will end.[84] And, ultimately, her mother's death is never narrated; Stella only encounters her mother's body in the street when a policeman asks, "What happened to the old trumpet?"[85]

The effect of these anachronies is to deprive the narrative of any sense of eventfulness. As we are pitched backward and forward from past to future, the narrative present in which events take place recedes before us. Lesley Marx has identified the time of Hawkes's novel as "ghostly," "voiding the present" in a "vertiginous spiraling of analepsis and prolepsis."[86] This has the effect of denying the possibility of any temporalized movement at all, evacuating the narrative of any presence whatsoever.[87] The novel's central event, Zizendorf's assassination of Leevey, is also narrated in this empty time, as the text presents it both too early and too late but never right on time. Zizendorf and his co-conspirators wait for Leevey by the side of the road; the problem is that he has fallen behind schedule, and his would-be assassins can no longer adhere to their plan. The only alternative, then, is an indefinite delay:

> "He's late," said Fegelein.
> "Yes."
> "No sleep for us then."
> "Wait, have patience," I answered.[88]

As their waiting becomes interminable, the future disappears as a horizon of temporality:

> "I don't want to go forward tonight; you mustn't make me..."

"Stop that. You know there isn't any forward."[89]

And then, at the moment when Leevey is to ride by, the narrative jumps back to earlier in 1945, to the riot at the asylum on the hill above *Spitzen-on-the-Dein*. This lacuna itself defers our anticipation; but when we return to the narrative present, it is already too late: "Leevey was killed outright when his motorcycle crashed into the log. He was pitched forward and down into an empty stretch of concrete."[90] We have, in fact, had a flash of the event earlier when Herr Stintz, a minor character, watches from an extreme distance as "the light [of Leevey's motorcycle] flared once and went out."[91] We get the event both before and after it happens, and thus, there is no narrative tension to his death at all: "He fell as easy as a duck, that area commander... Nothing grand about the commander's end at all!"[92] "There isn't any forward" because there isn't any present from which to depart. Hawkes's text becomes almost entirely populated by the twin ellipses of the elided event and the anachrony of its revelation. All we are left with is a continually emptied present. In constantly moving forward and backward, the text destroys the possibility of temporal movement altogether. Thus, the cornerstone event of the assassination, which promises to inaugurate an epochal revolution in *Spitzen-on-the-Dein* and to motivate the narrative of *The Cannibal*, is reduced to an absence, a site where past and future cross paths without conjoining. This is why, when asked by his step-daughter in the novel's final lines, "Has anything happened?" Zizendorf can reply, "Nothing. Draw those blinds and go back to sleep."[93]

To this extent, Hawkes is able to perform the temporality of the nuclear age in ways that Hersey cannot. Both texts suggest, however, that it might be impossible to capture the nuclear threat in the "right" way, and thus that allegory might function as a preferred, perhaps even necessary, trope to represent its "unthinkable" specter of total annihilation.[94] And, as I have been contending, both texts adumbrate a broad trajectory for interrogating a field of nuclear modernism. Hersey and Hawkes each demonstrate a belief that, for however the atomic bomb has disrupted the calculation of historical time, it has only prompted the emergence of a different form of temporality that is equally knowable and representable. Totality remains a viable option for them; literature still offers the possibility of grasping, if not redeeming, the nightmare of history. Hersey thus attempts to fold the nuclear back in to a progressive modern linearity, repressing its essential futurelessness, while Hawkes is more willing to experiment with the temporal possibilities that the nuclear age affords for literary practice. But the point is that both take the threat of nuclear annihilation seriously, and neither has abandoned a modernist project of formal novelty or authentic self-expression. The absurdity of gargantuan stockpiles of weapons thousands of times more destructive than Hiroshima was not yet in evidence, and the default response to that situation was not citational irony or fragmentary pastiche. That project may reach its own limits and encounter a certain historical futility, but its failure is not self-conscious, foreordained, or presumed. Rather, treating late modernism as a phenomenon inflected by the nuclear age revises many of the ordinary assumptions of twentieth-century periodization: it brings the modern and the postmodern much closer together, allowing us to imagine their contiguity while still recognizing a crucial mid-century hinge. That hinge allows one to articulate a different genealogy of postmodernism, emerging not only, as Andreas Huyssen has contended, from a dissatisfaction with an increasingly domesticated and politically impotent modernism,[95] but also as the product of a thwarted interrogation of historical time. Inasmuch as "it is safest to grasp the postmodern as an attempt to think the present historically in an age that has forgotten how to think historically in the first place," recognizing the place of late modernism allows us to grasp the final convulsions of historical memory.[96] That is, nuclear modernism might be understood as an attempt to answer the question, "If the atomic bomb brings to a close the historicity of modern endless progress, then what does the form and movement of historical time look like in the nuclear age?" When that question proved unanswerable, postmodernism, then, would be the abandonment of its terms altogether. In "Periodizing the American Century," Ann Douglas identifies two distinct moments within what she overarchingly terms "postmodernism," which I must quote here at length:

> I see the half century since World War II as divided into two stages, the first lasting from roughly 1945 through the early 1960s, the second running from the mid-1960s until the present. The latter is the period Anthony Appiah has labeled "post-optimism," when, in Masao Miyoshi's words, "the return to 'authenticity' is a closed route," and, to quote Edward Said, "innocence is... out of the question." The first generation of post-World War II artists in both the First and the Third Worlds, faced with the psychotic behavior and elaborately systematic deceit of the cold war era, were nerved to fresh acts of resistance and self-expression. These desperately creative acts of heroic subjectivity were attempts at what Jack Kerouac called "100 percent personal honesty," a romantic reinvention of charisma designed to declassify every kind of information for revolutionary political and artistic ends. In this earlier period, art and thought still seem grounded in specific geographical places and historical times; new forms, political and artistic, are believed to be possible. There is an outside to the system, a place where protest is meaningful and

consequential... To me, the second, "postoptimism" stage of the culture of late capitalism is the less interesting one. Despite its achievements, the feminist movement chief among them (the early cold war era was a time of masculine revolt), I see it as the time when many of us lost our way. At its worst, this stage was little better than a quagmire of intellectual and political compromise, of overinvestment in pastiche and irony, a cultural moment ashamed of its hopes and defaulting on its dreams.[97]

Douglas's key political coordinates for understanding the Cold War are significantly different than the ones I have proposed here (she privileges the secrecy of the national security state over the atomic bomb it supposedly guarded),[98] but fundamentally I find this periodization to mark the distinction I have been attempting to elaborate. I would only insist, however, that we see more clearly and deeper into Douglas's first "stage" by viewing it under the microscope of nuclear modernism; the difference is more than semantic. The alignment of Nuclear Criticism with the postmodern has foreclosed the asking of the "optimistic" questions about historical time and historical change. Inserting nuclear modernism into our periodization of the nuclear age, and into the practices of Nuclear Criticism, holds open the possibility of asking them again, anew, from a distinct perspective.

Notes

[1] Anne O'Hare McCormick, "The Promethean Role of the United States," *New York Times*, 8 August 1945, 22.
[2] Paul Boyer, *By the Bomb's Early Light: American Thought and Culture at the Dawn of the Atomic Age* (Chapel Hill: University of North Carolina Press, 1994), xxi.
[3] William H. Lawrence, "2D Big Aerial Blow," *New York Times,* 9 August 1945, 1.
[4] McCormick, "Promethean Role," 22.
[5] Quoted in Boyer, *Bomb's Early Light*, 5.
[6] The expression, of course, belongs to Frank Kermode, *The Sense of Ending: Studies in the Theory of Fiction* (Oxford University Press, 1968), a text itself very much concerned with how the nuclear imaginary affects the temporalities of narrative.
[7] Barnet Nover, "To What Fell Use," *Washington Post*, 9 August 1945, 6.
[8] Spencer Weart, *Nuclear Fear: A History of Images* (Cambridge: Harvard University Press, 1988), 106.
[9] Sarah Henstra, *The Counter-Memorial Impulse in Twentieth-Century English Fiction* (New York: Palgrave Macmillan, 2009), 80.

[10] I use this term in a slightly looser manner than that pioneered by Richard Klein, who, in "Proposal for a *diacritics* Colloquium on Nuclear Criticism," *diacritics* 14, no. 2 (Summer 1984): 2-3, and "The Future of Nuclear Criticism," *Yale French Studies* 77 (1990): 76-100, has used it to describe a particular critical methodology. My use is closer to that of Daniel Cordle, who designates a post-Cold War "second phase" of Nuclear Criticism no longer concerned exclusively with preventing nuclear war, and thus capable of exploring a wider range of atomic age cultural phenomena. See Daniel Cordle, *States of Suspense: The Nuclear Age, Postmodernism and United States Fiction and Prose* (Manchester University Press, 2008), 23-5.

[11] While a complete list would be impossible to construct, some of the notable articles in the field include Tony Jackson, "Postmodernism, Narrative, and the Cold War Sense of an Ending," *Narrative* 8, no. 3 (October 2000): 324-38; Rob Wilson, "Postmodern as Post-Nuclear: Landscape as Nuclear Grid," in *Ethics/Aesthetics: Post-Modern Positions*, ed. Robert Merrill (Washington, D.C.: Maisonneuve Press, 1988), 169-92; and Ann Douglas, "Periodizing the American Century: Modernism, Postmodernism, and Postcolonialism in the Cold War Context," *Modernism/Modernity* 5, no. 3 (1998): 71-98. Other full-length studies that rely on this periodizing assumption are: Walter Davis, *Deracination: History, Hiroshima, and the Tragic Imperative* (Albany: State University of New York Press, 2001); J. Fisher Solomon, *Discourse and Reference in the Nuclear Age* (Norman: University of Oklahoma Press, 1988); Peter Schwenger, *Letter Bomb: Nuclear Holocaust and the Exploding Word* (Baltimore: The Johns Hopkins University Press, 1992); and Tobin Siebers, *Cold War Criticism and Politics of Skepticism* (Oxford University Press, 1993).

[12] Jackson, "Postmodernism," 327.

[13] Wilson, "Postmodern as Post-Nuclear: Landscape as Nuclear Grid."

[14] For a more thorough explanation, see Fredric Jameson, *A Singular Modernity: Essay on the Ontology of the Present* (London: Verso, 2002), esp. 99-105.

[15] Brian McHale, *Postmodernist Fiction* (London: Routledge, 1987), 6.

[16] Ibid., 9-10.

[17] Fredric Jameson, *Postmodernism, or, The Cultural Logic of Late Capitalism* (Durham, NC: Duke University Press, 1991), 305.

[18] Henstra, *The Counter-Memorial Impulse in Twentieth-Century English Fiction*, 80.

[19] I am attempting to read the "temporal dominant" of Jameson's determination of modernity in the dual exigencies of the always new and the colonization of the non-modern. It would seem to me that both projects—the production of ever-new forms and the continuing abolition of nature, the countryside, older or simply other modes of production, etc.—demand an aleatory conception of the future, even to the extent that such a future entails a particular teleology. That is, the demands for novelty and agglomerative totalization are symptomatic of the modern historical temporality of endless progress, as opposed to Christian rectilinearity or ancient cyclicality. To be sure, such endlessness does not mean that early twentieth-century modernism did not have its own eschatological preoccupations, but it

distinctly viewed them from the perspective of survival and redemption; the decadence of the present could be endured and aesthetically transformed (Eliot is exemplary here). Utopia would be the converse corollary of this project, but would equally position futurity as its essential temporal horizon. On the "temporal dominant" of modernity, see Fredric Jameson, "The End of Temporality," *Critical Inquiry* 29 (Summer 2003): 695-718; on the frustrated desire of utopia, see Fredric Jameson, "The Politics of Utopia," *New Left Review* 25 (January-February 2004): 35–54.

[20] Jean Baudrillard, "What Are You Doing after the Orgy?," trans. Lisa Liebmann, *Artforum* 22, no. 2 (October 1983): 42-46.

[21] James Agee, "The Bomb," *Time*, 20 August 1945, accessed February 18, 2010, http://www.time.com/time/magazine/article/0,9171,797639,00.html.

[22] Klein's work is deeply indebted to Jacques Derrida, "No Apocalypse, Not Now (full speed ahead, seven missiles, seven missives)," trans. Catherine Porter and Philip Lewis, *diacritics* 14, no. 2 (Summer 1984): 20-31.

[23] Klein, "Future," 76, 81.

[24] Ibid., emphasis added.

[25] Jameson, *Postmodernism*, xi.

[26] Ibid., ix; Derrida, "No Apocalypse, Not Now," 23.

[27] Jameson, *Postmodernism*, xvi.

[28] Robert Jay Lifton and Greg Mitchell, *Hiroshima in America: Fifty Years of Denial* (New York: G.P. Putnam's Sons, 1995), 338-9.

[29] Paul de Man, "The Rhetoric of Temporality," in *Blindness and Insight: Essays in the Rhetoric of Contemporary Criticism* (Minneapolis: University of Minnesota Press, 1983), 207.

[30] Fredric Jameson, *The Political Unconscious: Narrative as a Socially Symbolic Act* (Ithaca: Cornell University Press, 1981), 29-30.

[31] On the blurring of victims and victimizers, see Robert Jay Lifton and Eric Markusen, *The Genocidal Mentality: Nazi Holocaust and Nuclear Threat* (New York: Basic Books, 1990), esp. ch. 8.

[32] For this history, see David Sanders, *John Hersey* (New York: Twayne Publishers, 1967), esp. ch. 2; and Michael J. Yavenditti, "John Hersey and the American Conscience: The Reception of 'Hiroshima,'" *Pacific Historical Review* 43, no. 1 (February 1974): 24-49.

[33] Yavenditti, "American Conscience," 42, 48.

[34] Weart, *Nuclear Fear*, 108.

[35] Mary McCarthy, "A Letter to the Editor of *Politics*," in *On the Contrary: Articles of Belief, 1946-1961* (New York: Octagon Books, 1976), 3.

[36] Ibid., 4.

[37] Ibid.

[38] Kingsley Widmer, "American Apocalypse: Notes on the Bomb and the Failure of Imagination," in *The Forties: Fiction, Poetry, Drama*, ed. Warren French (Deland, Fl.: Everett/Edwards, 1969), 142.

[39] Widmer, "American Apocalypse," 143.

[40] Ibid., 141.

[41] John Hersey, "The Novel of Contemporary History," *The Atlantic*, November 1949, 84, 82.
[42] Hersey acknowledged his indebtedness to modernists like Wilder, Dos Passos, Malraux, and Faulkner in an interview with Jonathan Dee, "The Art of Fiction 92: John Hersey," *The Paris Review* 100 (Summer-Fall 1986): 211-49. For a reading of Hiroshima that traces the influence of T.S. Eliot's *The Waste Land*, see Patrick Sharp, "From Yellow Peril to Japanese Wasteland: John Hersey's 'Hiroshima,'" *Twentieth Century Literature* 46, no. 4 (2000): 434–52.
[43] Alan Nadel, *Containment Culture: American Narratives, Postmodernism, and the Atomic Age* (Durham: Duke University Press, 1995), 3.
[44] Ibid., 53.
[45] Ibid., 60.
[46] Ibid., 60.
[47] Ibid., 65.
[48] Dee, "Art of Fiction," 228.
[49] Samuel Beckett, *Nohow On: Company, Ill Seen Ill Said, Worstward Ho* (London: Calder, 1989), 101.
[50] Nadel, *Containment Culture*, 53; Dee, "Art of Fiction," 228.
[51] Nadel, *Containment Culture*, 53, emphasis added.
[52] Tyrus Miller, *Late Modernism: Politics, Fiction, and the Arts Between the World Wars* (Berkeley: University of California Press, 1999), 12.
[53] Andrew Hoberek, "Cold War Culture to Fifties Culture," *Minnesota Review* 55-57 (2002): 143.
[54] This assessment is based on the research of Robert Lifton who, in *Death in Life: Survivors of Hiroshima* (New York: Random House, 1968), determines through interviews with survivors that the experience of the bombing predominantly takes the form of a "permanent encounter with death" and a feeling that "the whole world was dying" (30, 22). As I will explain below, this apocalyptic experience of the event is notably absent from Hersey's account.
[55] Hersey's attempt at temporal restoration certainly has affinities with similar gestures readily found in high modernist classics like Joyce's *Ulysses* (1922), which coincides narrative and temporal closure by containing the narration within a precise twenty-four hour cycle (marked textually by the book's opening and closing "s"). Diurnal plots are likewise found in Woolf's *Mrs. Dalloway* (1925) and *Between the Acts* (1941), suggesting a certain tendency to use natural temporal cyclicality as a way of enclosing the temporal ruptures enacted within the narrative.
[56] John Hersey, *Hiroshima* (New York: Vintage, 1989), 1.
[57] U.S. Strategic Bombing Survey, *The Effects of the Atomic Bombings of Hiroshima and Nagasaki* (Washington, D.C.: U.S. Government Printing Office, 1946), 3.
[58] For an extended discussion of this, again see Sharp, "From Yellow Peril to Japanese Waste Land."
[59] Hersey, *Hiroshima*, 4-5.

[60] Walter Benjamin, "Theses on the Philosophy of History," *Illuminations*, trans. Harry Zohn (New York: Schocken, 1969), 261.
[61] Hersey, *Hiroshima*, 43.
[62] Ibid., 45-9.
[63] Ibid., 17, 21, 42, 49, 56, 57, 66.
[64] Ibid., 16.
[65] "To Our Readers," *The New Yorker*, 31 August 1946, 15.
[66] Hersey, *Hiroshima*, 94, 98, 106, 106, 107, 131.
[67] Ibid., 152.
[68] John Hawkes, *The Cannibal* (New York: New Directions, 1962), 18 and *passim*.
[69] Robert Lifton, in *Death in Life*, observes that "olfactory imagery"—the peculiar smell of nuclear fission, burnt flesh, and eventually decaying bodies—plays a crucial role in how the world-ending quality of the bombing was experienced, and later in the survivors' psychic closing-off of the event. See esp. 23, 33-4.
[70] Hawkes, *The Cannibal*, 125.
[71] Ibid., 161, 10.
[72] Ibid., 9, 6.
[73] Ibid., 6, 37, 10.
[74] See Joan Bischoff, "John Hawkes' Horses of the Apocalypse," *Notes on Contemporary Literature* 6, no. 4 (September 1976): 12-14.
[75] Patrick O'Donnell, *John Hawkes* (Boston: G.K. Hall and Co., 1982), 31.
[76] Albert Guerard, introduction to *The Cannibal*, by John Hawkes (New York: New Directions, 1962), xvi.
[77] Peter Brooks, *Reading for the Plot: Design and Intention in Narrative* (Cambridge: Harvard University Press, 1992), 23.
[78] Gérard Genette, *Narrative Discourse: An Essay in Method*, trans. Jane E. Lewin (Ithaca: Cornell University Press, 1980), 64.
[79] Ibid., 33.
[80] Hawkes, *The Cannibal*, 12.
[81] Ibid., 47.
[82] Ibid., 13.
[83] Ibid., 80, 81-2.
[84] Ibid., 78.
[85] Ibid., 79.
[86] Lesley Marx, *Crystals out of Chaos: John Hawkes and the Shapes of Apocalypse* (Madison: Fairleigh Dickinson University Press, 1997), 38-9.
[87] Ibid., 43-4.
[88] Hawkes, *The Cannibal*, 126.
[89] Ibid., 127.
[90] Ibid., 157.
[91] Ibid., 143.
[92] Ibid., 169-70.
[93] Ibid., 195.

[94] On the relationship between the "unthinkable" nature of nuclear war and literature, see Peter Schwenger, "Writing the Unthinkable," *Critical Inquiry* 13, no. 1 (Autumn 1986): 33-48.

[95] See Andreas Huyssen, *After the Great Divide: Modernism, Mass Culture, Postmodernism* (Bloomington: Indiana University Press, 1986), esp. ch. 10.

[96] Jameson, *Postmodernism*, ix.

[97] Douglas, "Periodizing the American Century," 83-4.

[98] This distinction is more than a quibble over the relative importance of different historical agents and their causal forces; I would contend that Douglas's national security state belongs to a fundamentally dialectical Cold War logic of "us vs. them," in which the United States and the Soviet Union imagine themselves in relation to one another, whereas thinking about the atomic bomb is essentially anti-dialectical in that it both threatens to bring history to a halt and that the U.S. and the U.S.S.R. stand in the same relation to it, "us *and* them vs. it." Spelling this argument out is beyond the scope of this essay, but it does illustrate why it might be useful to distinguish between Cold War and Nuclear Criticism.

CHAPTER NINE

WAR AS PEACE:
AFTERLIVES OF NUCLEAR WAR
IN DAVID FOSTER WALLACE'S *INFINITE JEST*

JESSICA HURLEY

> In order that a war be a war, it has to have a form, otherwise it turns into
> a civil war, a war against oneself.
> —Paul Virilio, *Pure War*

It was touch and go for a while there. In the year leading up to the spring of 1984, when Nuclear Criticism was officially initiated at a conference at Cornell, Nena's nuclear protest song "99 Luftballons" was at Number 2 in the American charts, nuclear-scenario dramas *The Day After, Testament* and *Threads* were aired on television, and the Armageddon-themed Hollywood blockbusters *WarGames* and *The Terminator* were released. Nuclear war was in the air, on the radio, on the screen. Nuclear Armageddon seemed unavoidable, the vanishing point to which the vectors of history seemed to be inescapably drawn; Jonathan Schell, in his 1982 bestseller *The Fate of the Earth*, described living with the nuclear threat as "the cramped, claustrophobic isolation of a doomed present."[1]

Trying to ascertain what time we were living in became a pressing concern for Nuclear Critics. The great promise of Nuclear Criticism, Richard Klein argued in later years, was to re-conceptualize our attitude towards time in the face of the possibility that there would be no future at all, thus freeing us from the sense of history-as-inevitability which is produced by a future anterior model of history. Such a model is impossible in the face of there being no future to look back from: "Nuclear Criticism denies itself that posthumous, apocalyptic perspective, with its pathos, its revelations, and its implicit reassurances; it supposes that the only future may be the one we project forward from the time when total nuclear war, for the time being, has not taken place."[2] However, many of the articles from the 1984 volume of *diacritics* which sprang from the Cornell conference speak to the impossibility of such a mission rather than modeling a successful approach to this strange but promising nuclear temporality. Thinking about Armageddon in the future, even if to try and imagine it destroying the possibility of a future, inevitably leads to a future-oriented stance; as Dean MacCannell writes, "[Thinking] about the 'Bomb' always has taken the form of attempts to bring an imagined future moment, a moment of awful destruction, to bear upon the present."[3] This is, politically, not good: nuclear apocalypse scenarios, whether in the cultural milieu outlined above or in the political debates of the time, are "ways that thought has been deformed exactly so as to speed up the moment of destruction: we are all waiting for it now."[4] By imagining how Armageddon will look, we transport it imaginatively into our own time, then feel a certain anticlimax when it fails to materialize. Even Derrida, whose article steps very carefully around the dangers of "the anticipatory assimilation of that unanticipatable entirely-other,"[5] relies on a proleptic relationship of the present to the future, the gesture of imagining the unimaginable even while acknowledging that it cannot be imagined, to theorize the nuclear as literature's ultimate referent.[6] In 1984 the end of the world seems close, and getting closer with every attempt to reason or imagine it away.

There is, however, a form of Nuclear Criticism which argues that it is possible to imagine the nuclear future in an impossible way rather than as being impossible. Elsewhere in the Nuclear Criticism volume of *diacritics*, Derrick De Kerckhove argues that nuclear prolepsis does not take place in the imagination at all, but in fact:

> Consciously and also unwittingly, we have been rehearsing the nuclear blast ever since 1945. It may be that the taste for violence which makes up the content of most of our entertainment is itself symptomatic of the violence we do ourselves as we repress the very thought of the nuclear holocaust. Because of the bomb, we are all experiencing the psychological conditions of wartime permanently... Indeed, we are mutants before rather than after the nuclear blast.[7]

No future: no waiting / no waiting: no future. The nuclear threat is not a threat; it exists not in the future but in the present. The worst has already happened: anything worse than this and we won't know about it, being nothing but white shadows on the walls. This is a very specific form of eschatological time. The eschaton is the time between the Incarnation and the fulfillment of this event in the Apocalypse; nuclear eschatology exists after the incarnation of the bomb but in the impossibility of the fulfillment of its use (or of our knowing of its use, which as Derrida points out amounts to the same thing). The white shadows which would be left by the "1 Big 1"[8] become what Orin Incandenza in *Infinite Jest* wakes up to every morning: "His own impression sweated darkly into the bed beneath him, slowly drying all day to a white salty outline just slightly off from the week's other faint dried outlines, so his fetal-shaped fossilized image is fanned out across his side of the bed like a deck of cards, just overlapping, like an acid trail or timed exposure."[9] Or like a body which faces nuclear obliteration every day but which remains undestroyed, which continues to sleep, to wake, to live in cycles, but which leaves its faint shadow of the nuclear trace in its cyclical wake.

For Derrida, literature "exists in a future anterior relation to its own effacement... not just loss but the loss of loss, the termination of the capacity to symbolize, circumscribe and thus recover from loss, the loss of the very aftermath of loss."[10] *Infinite Jest* models a different kind of nuclear literature: a literature that exists not in relation to its own future destruction or a melancholic inability to mourn, "in the very possibility of its effacement," but which rather depends upon an understanding that in fact nothing will be effaced, that the worst has already happened, that the will-have-been of the future anterior is exactly the same as the how-it-is of the present tense.[11] By the time we get to the publication of *Infinite Jest* in 1996, the Berlin Wall is down, the Soviet Union has collapsed, and the nuclear condition is not the mutually assured destruction of two superpowers (and thus the whole world) but the proliferation of smaller-scale nuclear weapons: the globalization, if you like, of a lethal but not omnicidal technology. The planet has not been destroyed; the Armageddon which was irradiating popular culture in 1984 becomes the unstoppable, almost-invisible flow of something like nuclear capital in the vision of nuclear material passing through a "Terrorist Arms Bazaar" in 1997's *Tomorrow Never Dies*.[12]

Between then (1984) and then (1996) David Foster Wallace engages, at least metaphorically, with the relationship between nuclear war and literature. In the summer of 1993, Volume 13, Issue 2 of the *Review of Contemporary Fiction* is released, featuring a 24-page interview with David Foster Wallace conducted by Larry MacCaffery. This interview is beloved of Wallace's (surprisingly few) scholarly interlocutors, most of whom extend Wallace's thoughts on the limits of metafictional irony in a post-postmodern world into an analysis of his own style, generally ending up in the vicinity of what Paul Giles dubs Wallace's "sentimental posthumanism," which is "a more affective version of posthumanism, where the kind of flattened postmodern vistas familiar from the works of, say, Don DeLillo are crossed with a more traditional investment in human emotion and sentiment."[13] In the context of this paper, however, it is worth looking more closely at the language used by Wallace to describe literature's shift to the meta:

> It's almost like postmodernism is fiction's fall from Biblical grace. Fiction becomes *conscious* of itself in a way it had never been. Here's a really pretentious bit of pop analysis for you: I think you can see Cameron's *Terminator* movies as a metaphor for all literary art after Roland Barthes, viz., the movies' premise that the Cyberdyne NORAD computer becomes conscious of itself as *conscious*, as having interests and an agenda; the Cyberdyne becomes literally self-referential, and it's no accident that the result of this is nuclear war, Armageddon.
>
> Metafiction's real end has always been Armageddon. Art's reflection on itself is terminal.[14]

Wallace locates himself and his art, positioning it as he does as a response to metafiction and thus postdating it, as occurring after the end of time. Wallace is most closely aligned with De Kerckhove here, in that there seems to be an assimilation of the nuclear into the everyday; but where De Kerckhove maintains a forward momentum ("The bomb is not obsolete yet because it has not finished its job"),[15] Wallace's nuclear language settles for a combination of circularity (the loop of Cyberdyne's self-aware consciousness) and intensification: "TV didn't invent our aesthetic childishness any more than the Manhattan Project invented aggression. Nuclear weapons have simply intensified the consequences of our tendencies, upped the stakes."[16] Here, at the end of the Cold War, Wallace's nuclear imagery seems to accept that the bomb itself, the real life bomb, will never fall; it does not exist in the future at all but rather lives in the interstices of the present, intensifying aspects of it and closing off access to any thought of a future, leaving us in a time that is intensely circular without repetition, terminal without stopping, as in the terminal of a battery, a point of flow.

The New Colonialism: Forms of War in *Infinite Jest*

This subsumption of Armageddon into everyday life produces a distinct model of war in *Infinite Jest*, where politics and diplomacy become indistinguishable from war and violence. The novel is set in a near future in which America, led by ex-lounge singer President Gentle and his Clean United States Party, has taken a radical approach to waste management. This involves signing treaties with Mexico and Canada to become the Organization of North American Nations (O.N.A.N.), while deliberately poisoning most of New England with toxic waste. In the admirable summary of N. Katherine Hayles:

> It soon becomes apparent that Gentle's version of "Interdependence" is merely rampant nationalism under another guise, for the U.S. uses its O.N.A.N. clout to force Canada to annex the toxic waste area so it will not soil U.S. cleanliness. Surrounded on the south by Lucite walls and fed by great fans intended to keep toxic fumes from escaping, "The Great Concavity" as it is called in the U.S. (from the Canadian perspective it is "The Great Convexity") signals the transition from imperialism to experialism. While imperialism is about expropriating valuable natural resources from less powerful nations, experialism is about forcing them to accept the industrial wastes that result when the expropriated natural resources are turned into capitalist commodities.[17]

This piece of geopolitical prestidigitation is typical of the way that war and violence operate in *Infinite Jest*. It is a political process which has the results of a war: Canada is basically annexed to America (the signing of the O.N.A.N. Treaty is referred to twice as "Findlandization" and four times as a continental "Anschluss"), maintaining just enough autonomy for the U.S. to pretend that the Concavity has nothing to do with them. Carl von Clausewitz defines war as *"an act of violence intended to compel our opponent to fulfill our will."*[18] Here we have the will of one party fulfilled, but using a kind of violence that is far from Clausewitz's original referent; it is closer to his most famous statement: "War is nothing but a continuation of political intercourse, with a mixture of other means."[19]

Neither a cold war nor a hot one, this is more of a simmering political situation with no end in sight: an intensification, to recall the MacCaffery interview, of the political situation rather than a shift into a traditional War. While the plot summary above may make it seem like this scenario has nothing in common with traditional narratives of nuclear destruction, the consequences of what is officially called, under the Gentle administration, "Territorial Reconstruction," are distinctly nuclear in their presentation. The Concavity is closed to the public not only because it is lethally toxic but because of the mutated creatures which live there: giant feral hamsters, descended from pets abandoned during the forced evacuation of what was then New England, roam in packs, and there are rumors of equally outsized babies which are "anomalous and huge, grow but do not develop, feed on the abundance of annularly available edibles the overgrowth periods in the region develop, do deposit titanically outsized scat, and presumably do crawl thunderously about, occasionally sallying south of murated retention lines and into populated areas of New New England."[20] Similarities between these mutated monsters and the post-World War II horror movies of the 1950s which also played on fears of radiation and the nuclear (*Godzilla, King of the Monsters*; *Them!*; *Attack of the Giant Leeches*) are pointed up by the language used to describe the feral hamsters, which is distinctly reminiscent of U.S. government-produced nuclear attack safety films such as *Duck and Cover* or *Survival Under Atomic Attack* (both 1951):

> With respect to a herd of this size, please exercise the sort of common sense that come to think of it would keep your thinking man out of the southwest Concavity anyway. Feral hamsters are not pets. They mean business. Wide berth advised. Carry nothing even remotely vegetablish if in the path of a feral herd. If in the path of such a herd, move quickly and calmly in a direction perpendicular to their own. If American, north not advisable. Move south, calmly and in all haste, toward some border metropolis—Rome NNY or Glens Falls NNY or Beverly MA, say, or those bordered points between them.[21]

Giant fans blow the toxic waste away from the American border towards Québec, resulting in babies being born with massive congenital deformities, the most consistent of which is having no skull. The result of these "paper atrocities" is a massively violent insurgency of Québecois separatist radicals, the most extreme of whom, in one of the novel's main plot lines, want to destroy the U.S. from the inside by distributing a film (entitled *Infinite Jest*) so addictively pleasurable that no one who watches it can survive.[22]

Here, then, we have the nuclear recast into the temporality of what Rob Nixon calls "slow violence." Nixon uses this term to describe processes such as "toxic buildup, massing greenhouse gases and desertification" which kill millions of people worldwide each year without these deaths having the impact of more dramatic, immediate acts of violence: "[Politically] and emotionally, different kinds of disaster have different heft. Falling bodies, burning towers, exploding heads have a visceral, page-turning potency that tales of slow violence cannot match."[23] *Infinite Jest*, however, complicates Nixon's distinction. Even the violence engaged in by the Québecois terrorist group *Les Assassins des Fauteuils Rollents* or AFR, which includes such spectacular acts as planting mirrors across American highway passes so that drivers drive off of the road and explode, gets slowed down in its representational impact if not in the actual impact of car and cliff:

> For over twenty months the scores of burnt-out hulls piling up in Adirondack chasms were regarded as either suicides or inexplicable doze-behind-the-wheel-type single-car accidents by NNY State Troopers who

had to detach their chinstraps to scratch under their big brown hats over the mysterious sleepiness that seemed to afflict Adirondack motorists at what looked to be high-adrenaline mountaintop passes."[24]

Slowed down and stretched out (*IJ* weighing in at 1079 pages), violence in *Infinite Jest* manages to be both nuclear and slow at the same time, the nuclear as a process rather than as an event.

This emphasis on different speeds of war takes us back to the early 1980s when Paul Virilio theorized Pure War. Pure War is a specifically nuclear condition. Virilio writes that "[we] can no longer even speak of wars" because "[war] today is either nuclear war or nothing."[25] The actual declaration of war would be the end of the world in a nuclear age because of mutually assured destruction, so war becomes Pure War, entirely composed of deterrence, the ever-ratcheting steps taken to stay one logistical step ahead of the opponent which Michael McCanles calls "the cruelest of hoaxes: a deterrence that can guarantee security only by threatening total annihilation."[26] This is a logic that is pushed to an extreme in *Infinite Jest*. The most sustained recounting of the process of Territorial Reconstruction in *Infinite Jest* comes in the form of a claymation/puppet film which is described as "mixing real and fake news-summary cartridges, magazine articles, and historical headers from the last few great daily papers."[27] This postmodern approach to history makes it difficult to make claims about "what happened" in the fake history of the novel, but we are told that a "Pretty Obviously Homemade Subheader" in the film informs its viewers that Rodney Tine, head of the secret and scary Office of Unspecified Services in the Gentle administration, "THREATENS TO DETONATE UPSIDE DOWN MISSILES IN U.S. SILOS, IRRADIATE CANADA W/ AID OF ATHSCME HELL-FANS – Header; 'WILLING TO ELIMINATE OWN MAP OUT OF SHEER PIQUE' IF CANADA NIXES RECONFIGURATIVE TRANSFER OF 'AESTHETICALLY UNACCEPTABLE' TERRAIN."[28] Pushing "the remorseless logic of game theory" to its limits means that the state becomes openly, visibly suicidal; the opponent must believe that the state would genuinely be, for example, "better dead than Red."[29]

In the face of total annihilation, Virilio argues, "war is no longer in its execution, but in its preparation"[30]—a preparation which entails transferring more and more resources from the civilian population to the military, resulting in an underdevelopment of the State's own population in an act of what Virilio varyingly calls "endo-colonization" or "the suicide State."[31] *Infinite Jest* offers an interestingly globalized spin on this idea, in which America can only declare war by incorporating Canada into itself (or itself into Canada, depending on where you stand in the Concavity/Convexity debate) and then conducting an undeclared civil war against a part of its own territory. This differs from Virilio's conception of the suicide State, which is located specifically within the nation state: in order to think of a state acting against the interests of its own population, he takes as axiomatic the existence of states with clearly defined borders and clearly defined populations. In *Infinite Jest*, however, both geography and history (the two defining factors which go into the construction of the nation state) have become distinctly relativized.[32] How can one think of America or Canada as states when it is impossible to know what shape they are? Or, with any degree of certainty, what was said in the diplomatic meetings which define those borders? The very ability of America to dispose of its waste in what is officially Canadian territory suggests that an entirely different model of national sovereignty is at play here.

One World: The Global Nuclear

Virilio is still writing in the age of the threat of nuclear Armageddon; for him, in 1983, there is no alternative to Pure War between and within clearly defined nation states. Fortunately, however, in the series of interviews collected in the volume titled *Pure War,* Virilio and his scholarly comrade Sylvère Lotringer take exactly the journey that we are trying to trace here, as they return to the topic thirteen years after the initial publication of the book in 1983: "[Between] 1984 and 1996 we saw it all," Virilio writes, confidently.[33] In 1996, Virilio states that Pure War has moved into the realm of science, which now progresses solely for its own sake, and that deterrence has ceased to have any meaning in an age of nuclear proliferation.[34] From Pure War to what, then, in *Infinite Jest*?

In his summation of the "all" that he has seen between 1984 and 1996, Virilio lists "the Berlin Wall, the implosion of the Soviet Union and the Gulf war," as well as "Daboville's circumnavigation in a rowboat and the bombing of the World Trade Center."[35] He leaves out many things, obviously; the coming into force of NAFTA on January 1st, 1994, for example, or the establishment of the World Trade Organization on January 1st, 1995. This is not to say that he takes no interest in globalization. He touches upon it briefly in the 1996 interviews, where he senses more than describes how massive increases in the speed of human travel and communication have inscribed a certain limit around the world: "[We're] backed into a corner too, you see. All of us. Not simply a deathly impasse, but stuck in the corner as a result of globalization and everything we've been describing."[36] Not an impasse, which suggests a blockage in the path of a line, but a corner, equally blocked but this time in 3-D space with multiple, confusing and indeterminate vectors from which one might be threatened. In *Infinite Jest*, as we have seen, the suicidal nuclear threat leads to a form of violence which is both globalizing and ongoing: it shifts borders and makes them indeterminable, and those borders become liminal spaces where an ongoing nuclear catastrophe exists with no end in sight.

A useful way to think about globalization and indeterminacy is provided by Leerom Medovoi in her 2007 article "Global Society Must Be Defended: BIOPOLITICS WITHOUT BOUNDARIES." Medevoi mostly discusses the relationship between globalization and the war on terror, locating her project post-2001, but she offers useful readings of the 1990s in which she describes the hidden violent operations of globalization in order to demonstrate how the war on terror brings these secret maneuvers into the light. Medovoi describes how in

> the world of globalization in the 1990s... violence was always effaced by the need for regulation. For once, global exercises of biopower disavowed almost completely the military character of their project. Economic violence was merely 'structural adjustment,' every apparent war only a police action, every conflict with some 'way of life' a managing of risk to the global social body.[37]

We might usefully recall Virilio here: Pure War, the operation of the suicide State, is secret war, it is "not the kind which is declared."[38] As Remy Marathe, wheelchair-bound operative of the AFR in *Infinite Jest* tells a suicidal, depressed young marijuana addict in a bar that "our own...leaders have been subverted to pretend the invasion is alliance; we very few legless young cannot repel an invasion; we cannot even make our government admit that there is an invasion."[39] Indeterminacy of borders, a confusion between a state and not-state, is produced by globalization to enable "the collapse of liberal society's traditional distinction between the internal and external enemy, as well as between the practices by which each is targeted: regulation and warfare, respectively."[40] It is in the interests of the globalized state that these distinctions become indeterminate enough that its violence not be recognized as such, for it is part of the neoliberal ideology that states no longer apply violence. In *Infinite Jest* this process is laid out clearly, with an extra step: the need for violence against another state is precisely what causes globalization here.[41] Openly acknowledged international war (in the form of dumping nuclear waste on Canada) is not possible, so blur the boundaries, annex the neighboring country and then commit violence against it which is also yourself—but which will not be recognized or named as violence, ever.

This blurring of clearly defined categories extends beyond national borders in *Infinite Jest* into the temporal, and the nuclear is intricately linked to how regimes of time are both represented by and operative in the novel. In the near future of *Infinite Jest*, linear time has, at least officially, come to an end. Years are no longer numbered but are sponsored by corporations: Year of the Depend Adult Undergarment; Year of the Perdue Wonderchicken; Year of Dairy Products from the American Heartland. This is as a direct result of the Experialist coup. The immense cost of the forced migration of the inhabitants of New England after it has been poisoned by the waste products of annular fusion must be recouped by the government somehow, and selling time to corporate sponsors is offered as a solution. Annularization is what the nuclear has become when it is no longer officially weaponized. In the world of the novel, nuclear annihilation has become nuclear annulation, or annular fusion: an energy producing process (America is, in this world, energy-independent) by which toxic radioactive waste is fissioned and the products are fused, creating energy and more toxic radioactive waste, which is then used for the first stage in the process, and so on.[42] The invention of this process lies at the base of two out of the three main narratives in the novel: the massive amounts of toxic waste it produces are what creates the need for the Concavity and American Experialism, setting off the plot of the AFR's revenge scheme, while the inventor of some of the vital technology for the process uses the money he gets for it to establish the tennis academy where nearly half of the novel is set. Since the invention of annular fusion thus leads to the establishment of Subsidized Time, the non-linear temporality at work in the novel results directly from a new nuclear scenario. Indeed, nuclear annulation is consistently linked in *Infinite Jest* to new models of temporality. In the Concave space of the nuclear itself, time has become relative: radio personality Madame Psychosis has on the screen which hides her from view in the studio "four clocks set for different Zones, plus a numberless disk someone hung for a joke, to designate the annularized Great Concavity's No-Time."[43] And elsewhere, Michael Pemulis (teenaged Enfield Tennis Academy student, math whiz, and school drug dealer) explains to a confused younger student the "most basic kiddie-principles of the relativity of time in extreme organic environments."[44] Relativity of time, "time in flux," "accelerated phenomena, which is actually equivalent to an incredible *slowing down* of time;" all of these alternatives to linear time exist because of the shift from nuclear fission to annular fusion.[45]

    The primary importance of this shift from one kind of nuclear threat to another is emphasized by its position in the novel's *fabula*. Plot-wise, the story of *Infinite Jest* is kicked off with the invention of annular fusion. Aside from a dramatic monologue in the voice of James Incandenza's father set in 1960, in which nothing really happens, the earliest point in the story (which occurs just under halfway through the narrative) is a recollection by James Incandenza, the dead patriarch of the central-to-the-novel Incandenza family, set in 1963. James was molded by his dipsomaniacal father into a youthful tennis prodigy, went on to become an ocular physicist, founded the Enfield Tennis Academy, and spent the last years of his life making aggressively unpopular avant-garde films (the last of which is *Infinite Jest* or "The Entertainment") before committing suicide by putting his head in a microwave oven. James's story involves a deeply disturbing account of helping his parents take a bed apart, with his father passing out in his own vomit halfway through. While his mother is cleaning up the mess, the young James leaps onto his own bed, knocking over a lamp in the process which shears a brass knob from his dresser, which rolls in frictionless circles on the floor. As he watches, "[It] occurred to me that the movement of the amputated knob perfectly schematized what it would look like for someone to try to turn somersaults with one hand nailed to the floor. This was how I first became interested in the possibilities of annular fusion."[46] The combination of slapstick and extreme violence in this final image typify not only the style of the novel, but what happens to violence itself: it is no longer the cause of a teleological, *War and Peace*-style model of history moving on via battles, but of an annular or cyclical process which produces a similar model of history. Each revolution, in this case The Year of Glad or The Year of the Tucks Medicated Pad, passes through a full rotation, but these rotations seem to occupy the same space, with no sense of forward movement. The subsumption of the nuclear into a slow violence model creates in time what globalization creates in space: a circular (or spherical) enclosed system within which unlimited movement is possible but the outside of which is impossible to access.

The hermetically sealed, globalized form of space-time, which results from what we might call post-nuclear violence, is the biggest difference between Wallace's response to the nuclear threat after the Cold War, and that of Derrida and the other Nuclear Critics during it. The temporality of Derrida's thought in "No Apocalypse, Not Now" has been variously described as "the time of the interval,"[47] that strange moment of deconstruction which both accelerates and suspends time, and the "*pas encore.*"[48] What these approaches both describe is the sense of a temporality which keeps on going because the ending is not known, because it is, especially when thinking about the annihilation of the species and the archive, fundamentally unknowable. In *Infinite Jest*, however, the future is with us from the first page. The chronology of *Infinite Jest* starts at the end: the first chapter occurs at the latest point in the story of the novel, thirteen-months after the second latest point in the story. To be sure, there is narrative in the book: we flash backwards and forwards, we are never quite sure where we are (especially given the use of Subsidized Time), plot lines vector along through time and converge, or not; but there is no future here. We start at the chronological end, and everything we read after it is in the narratological past. Given how much the narrative jumps in time, it is surprising to realize that the opening section of the novel is its absolute temporal limit. We leap into the imaginary near-future of the novel, but once we are there, the formal structure of the novel prevents any further leaps.

To describe *Infinite Jest* as a closed circuit is to go against much of the extant criticism of the text. No less a critic than Fredric Jameson uses it as an exemplum of texts which "by fragmentation and imperfection or by a dizzying multiplication of presences on the page, somehow evade form and reification."[49] The temporality of the novel is more usually described as representing a kind of open-ended, addictive drive. Addiction and compulsion are major themes of the novel, and both Frank Cioffi and Daniel Carpenter Grausam lay out an equivalence between the addictions described in the novel and the experience of reading it (and are, as early critics, often referenced by other critics who agree with them).[50] For Cioffi, the alienation-effects of the paratext and esoteric vocabulary become addictive, while the narrative "keeps opening up interesting, urgent questions that the reader needs to have answered," and "[on] the microlevel of the sentence or paragraph, Wallace's turns of phrase produce the verbal equivalent of a 'fix'."[51] Grausam takes this analogy further:

> The readerly experience of *Infinite Jest* is a phenomenological analogue to the valorization of radical incompleteness that the novel's thematics offer. Taught to desire, the reader is akin to both the addict and the tennis prodigy, in that they learn the goal they have been taught to desire, and work so hard to achieve, is illusory at best, and fatal at worst, and that a model of experience as radically incomplete is the only satisfactory relationship to text and world.[52]

While I do not necessarily disagree with these analyses of the phenomenology of reading this bizarre and compulsion-inducing book,[53] it seems that the kind of radical incompleteness described by Grausam ignores the difference between the novel's themes and its form. This is a difference we should pay particular attention to in this text, not only because it is so formally challenging (its length, the 98 pages of footnotes), but because this very distinction is pointed out in the text itself. Again, you don't write a book that you need two bookmarks and special lumbar support to read unless you want to draw your reader's attention to form, but Wallace also drops a large hint by having James Incandenza's son Hal describe his work (and remember that this is the man who made the film *Infinite Jest*) as follows:

> I guess I could see some Séparitisteurs trying to read *The ONANtiad* or *Brick* as anti-Reconfiguration films. Maybe stuff like *Poultry in Motion*... And *Imminent Domain*'s allegedly on one level an allegory about the Concavity, though that overlooks that Gentle wasn't even President when that came out. But you can tell your Subject that Himself's work was all very self-consciously American. His interest in politics was subordinate to form. Always.[54]

While not wanting to draw overly simplistic connections between the author of the book and the great *auteur*-figure in the book, ignoring form altogether in the face of this seems equally rash. Take the footnotes, for example. *Infinite Jest* has 388 footnotes over 98 pages, ranging from a couple of words to the length of a goodly short story. For Grausam and Cioffi, the footnotes produce a moment of delay which feeds into the addictive nature of the text; the constant disruptions in the reading experience provide a moment of frustration which is then quickly satisfied. But they are also a strictly formal feature of the text: a paratext, a supplement. Like any supplement, this paratextual apparatus is added on to a text which is complete, "a plenitude enriching another plenitude"; it also, like any supplement, "adds only to replace. It intervenes or insinuates itself *in-the-place-of*; if it fills, it is as if one fills a void. If it represents and makes an image, it is by the anterior default of a presence."[55] The footnotes add something to the text, and in so doing, they draw attention to an originary lack in it.

Here, however, while I want to think with Derrida, Wallace, I suggest, thinks against him. The footnotes themselves have their own circularity: fn. 1 and fn. 388 are both explanations of a slang or medical term for drugs, for example. Some story elements exist solely in the footnotes, developing their own minor plot arcs and circles; many footnotes refer to other footnotes and some footnotes have their own footnotes. It would be going too far to say that the paratext could stand alone, but parts of it certainly could, which suggests that we are not meant to read the footnotes as pointing up an unfillable void in the main text. Rather, they point out a void in the text which only they can, and do, fill. This is particularly the case given the alternate-reality world that the novel is set in. Footnotes suggest an external referent, an explanation via a real-world third party which will clarify an initial confusion between the text and the reader. Here, however, because the world of the novel is so noticeably different from our world, the idea of referring to something in the "real" world stays strictly on an intradiagetic level. There is, literally, nothing outside the text; a footnote that recounts to us the story of Hal eating mold as a baby, or which explains to us the layout of Enfield Tennis Academy, remains resolutely fictional. Far from resisting form, *Infinite Jest* does everything it can to remain an object, fixed between its front and back covers.

If we accept, then, that *Infinite Jest* is both formally and thematically a closed loop, it cannot be working in the same temporal model as that of "No Apocalypse, Not Now," in which "[we] have absolute knowledge and we run the risk, precisely because of that, of not stopping"[56]—but in which, also, that absolute knowledge could mean either the wherewithal to build and explode a nuclear bomb or the wisdom not to do so, leaving the end of this piece caught between two equally necessary paths, and still moving forward ("I can't go on, I'll go on").[57] For Virilio, the closed loop is a symbol of the collapse of the distinction between spatial and temporal dimensions of globalization. In 1997, he describes the coming age of globalization as "the end of an era...the end of a temporal regime."[58] It is difficult to know what he thinks we are moving into, since he refers to both pre-and-post globalization time as "historical" time with predictably confusing consequences. Only when he returns again to the question in 2007 does he explain what this new time might be, and he does it in spatial terms:

> The closure of the world is a revelation. The revelation of finitude is an entirely new situation. This is not the apocalypse. Ideas of the end of the world are of no interest to me. I've said elsewhere: 'The end of the world is a concept without a future.' What this revelation reveals is that globalization is a finitude. Societies used to rely on local finitudes: frontiers, city ramparts, meta-cities, nations. Nowadays the frontier is the world; the world is finalized... This is not the end of the world, but the realization of its finitude.[59]

The end of the end of the world. Not in the impossibility of there being living witnesses, as in Derrida, but in the closing of the loop, the revelation of the finitude. This is not an apocalypse because nothing is unveiled; what is revealed is what was there all along. In terms of *Infinite Jest*: the opening chapter features Hal, who will be one of two main focalizers of the novel, unable to communicate, producing nothing but horror in those who see him try. The next time we see him, some fifteen months prior to this in the narrative, he is perfectly normal. In the two-month span of this plotline, we learn about a number of things Hal might come into contact with which could cause this effect: The Entertainment, the time-twisting drug DMZ, the mold he ate as a child. The last time we see him in the book, people think he is smiling when in fact he is sad. That's it. We never know what happened, how or why. There are countless theories out there, but none of them are more convincing than any of the others.[60] What is revealed over the course of the book is revealed, and what is not, will never be known. To look further into the future than the end of the book is to go back to the beginning, thirteen months ahead, then fifteen months back, and so on.

The global nuclear, then, the age of nuclear proliferation and the spread of global capital (both with scant regard for the traditional sovereign state), figures in this text as a fundamental closure, a claustrophobia not of Schell's doomed present, but of a present which tells us it will go on and on, like this, forever. Nicholas Royle describes Derrida's attitude to the future in terms of a kind of terror: "[T]he idea of the future, which is 'a sort of monstrosity' and which 'can only be anticipated in the form of an absolute danger' [*Of Grammatology* 5]. This monstrosity is most notoriously figured in Derrida's work in terms of the visual – for example, in terms of the 'glimpse' and 'glance' at the 'formless, mute, infant and terrifying form of monstrosity' [293] at the end of his essay 'Structure, Sign and Play in the Discourse of the Human Sciences'."[61] In *Infinite Jest,* we come face to face with this monster in the figure of the baby deformed by nuclear waste,

> a normal-size and unferal infant but totally without a skull, lying in a kind of raised platform or dais by the fireplace with its shapeless and deskulled head-region supported and, like (shuddering), contained in a sort of lidless plastic box, and its eyes were sunk way down in its face, which was the consistency of like quicksand, the face, and its nose concave and its mouth hanging out over either side of the boneless face, and the total head had like conformed to the inside of the containing box it was contained in, the head, and appeared roughly square in overall outline, the head, and the woman with the lei of gull-heads and other persons in costumes had ingested hallucinogens and drank mescal and ate the little worms in the mescal and had performed circled rituals around the box and platform around 2355h., worshipping the infant, or as they termed it simply The Infant, as if there were only One.[62]

The formless, mute, infant monstrosity no longer comes from outside; it cannot, for there is no more outside. It is here, in a box which both contains and displays; it is both reality and symbol. *Infinite Jest* presents us with a closed system which disguises itself as everything, as if there were only One, and we lack even the prospect of the nuclear apocalypse as a thought-experiment escape route. Derrida's unknowable future may sometimes seem frightening or monstrous, but compared to the everlasting present of the global nuclear, the prospect of anything which comes from an outside, even "someone from whose mouth 'a sharp double-bladed sword' was emerging"[63] is appealing, for he will be the one who is again able to say "'I am the first and the last'."[64] In a world whose continued existence depends upon the non-fulfillment of the nuclear prophecy, a new Nuclear Criticism will have to be able to think an escape from the plastic box of the global nuclear which is thoroughly anti-apocalyptic. There may be no way back from the nuclear condition, but a Nuclear Criticism that can live comfortably without the promise of ultimate revelation may be able to provide an alternative to the Hollywood ending of the nuclear story. Seeking an escape from Wallace's claustrophobic present we sit in the dark and watch mushroom clouds bloom over L.A., New York, London, Tokyo, until we find that we are not dreading it after all, but waiting for it. The task of the twenty-first century nuclear critic will be to teach us how to stop waiting.

Notes

[1] Jonathan Schell, *The Fate of the Earth, and, The Abolition* (Stanford, Calif: Stanford University Press, 2000), 172.
[2] Richard Klein, "The Future of Nuclear Criticism," *Yale French Studies* 97 (2000): 79.
[3] Dean MacCannell, "Baltimore in the Morning... After: On the Forms of Post-Nuclear Leadership," *diacritics: A Review of Contemporary Criticism* 14, no. 2 (1984): 34.
[4] Ibid.
[5] Jacques Derrida, Catherine Porter, and Philip Lewis, "No Apocalypse, Not Now (Full Speed Ahead, Seven Missiles, Seven Missives)," *diacritics: A Review of Contemporary Criticism* 14, no. 2 (1984): 23.
[6] Ibid., 28.
[7] Derrick de Kerckhove, "On Nuclear Communication," *diacritics: A Review of Contemporary Criticism* 14, no. 2 (1984): 78.
[8] Russell Hoban, *Riddley Walker*, Expanded ed. (Bloomington: Indiana University Press, 1998), 19.
[9] David Foster Wallace, *Infinite Jest: A Novel*, Back Bay 10th anniversary pbk. ed. (New York: Back Bay Books, 2006), 43.
[10] Derrida, Porter, and Lewis, "No Apocalypse, Not Now (Full Speed Ahead, Seven Missiles, Seven Missives)," 27.
[11] Ibid.

[12] Paul Williams, "Nuclear Criticism," in *The Routledge Companion to Science Fiction*, 1st ed., Routledge Literature Companions (London; New York: Routledge, 2009), 253. Or, closer to our time and my heart, the collapse of civilian/soldier/policeman/terrorist distinctions in the face of a nuclear threat which has also effaced the difference between its peacetime and wartime uses in the second, fourth and sixth seasons of *24*. See Slavoj Zizek, "The Depraved Heroes of 24 Are the Himmlers of Hollywood," *The Guardian*, 27 January 2006, http://www.guardian.co.uk/media/2006/jan/10/usnews.comment.

[13] Paul Giles, "Sentimental Posthumanism: David Foster Wallace," *Twentieth Century Literature: A Scholarly and Critical Journal* 53, no. 3 (2007): 330. For further examples of this focus on irony vs. sincerity see Iannis Goerlandt, "'Put the Book Down and Slowly Walk Away': Irony and David Foster Wallace's *Infinite Jest*," *Critique: Studies in Contemporary Fiction* 47, no. 3 (2006); Mary K. Holland, "'The Art's Heart's Purpose': Braving the Narcissistic Loop of David Foster Wallace's *Infinite Jest*," *Critique: Studies in Contemporary Fiction* 47, no. 3 (2006); Catherine Toal, "Corrections: Contemporary American Melancholy," *Journal of European Studies* 33, no. 34 (December 2003): 317-21.

[14] Larry McCaffery, "An Interview with David Foster Wallace," *The Review of Contemporary Fiction* 13, no. 2 (1993): 134.

[15] Kerckhove, "On Nuclear Communication," 80.

[16] McCaffery, "An Interview with David Foster Wallace," 129.

[17] N. Katherine Hayles, "The Illusion of Autonomy and the Fact of Recursivity: Virtual Ecologies, Entertainment, and Infinite Jest," *New Literary History: A Journal of Theory and Interpretation* 30, no. 3 (1999): 685.

[18] Carl von Clausewitz, *On War*, New & revised ed. (Harmondsworth: Penguin, 1968), 101. Emphasis in original.

[19] Ibid., 402.

[20] Wallace, *Infinite Jest*, 1056n.

[21] Ibid., 93. Compare, for example, this extract from *Survival Under Atomic Attack*, narrated by the marvelous Edward R. Murrow and accompanied by footage of an all-American family with adorable children (including a young boy in a tiny cowboy outfit) moving "calmly and in all haste" to their pre-prepared shelter: "In the case of an attack or test drill, the alert will be a warbling siren which will last for three minutes. Pull down the shades or blinds and close the drapes against flying glass. Turn off the burners of your gas or electric stove. Disconnect any heating elements such as electric irons, hotplates or bathroom heaters. Close all outside doors, but leave them unlocked. Turn off the gas or oil burners. Taking shelter may be a race against time, even when you have advance warning. But possibly there may be no time. An attack could come without warning" (transcription mine). *Survival Under Atomic Attack (1951)*, 2011, http://www.youtube.com/watch?v=sTJ0GB1ltAQ&feature=youtube_gdata_player.

[22] Wallace, *Infinite Jest*, 776.

[23] Rob Nixon, "Slow Violence," *Chronicle of Higher Education* 57, no. 40 (July 1, 2011): 14.

[24] Wallace, *Infinite Jest*, 312. Note, also, that the last thing you would see when driving into a reflected image of yourself driving at night would be a blinding light. It gets everywhere, this nuclear imagery.
[25] Paul Virilio, *Pure War: Twenty Five Years Later*, New and updated ed. (Los Angeles, CA: Semiotext(e), 2008), 39-40.
[26] Michael McCanles, "Machiavelli and the Paradoxes of Deterrence," *diacritics: A Review of Contemporary Criticism* 14, no. 2 (1984): 17.
[27] Wallace, *Infinite Jest*, 391.
[28] Ibid., 407.
[29] Ibid., 324.
[30] Virilio, *Pure War*, 104.
[31] Ibid., 107, 116.
[32] I take my working definition of the nation state here from Benedict Anderson, *Imagined Communities: Reflections on the Origin and Spread of Nationalism*, (London & New York: Verso, 2006) and Zygmunt Bauman, *In Search of Politics*, (Stanford: Stanford University Press, 1999). Both theorists describe the ways in which the nation is defined by geographical and historical boundaries which are complexly and often retroactively constructed in order to provide a locus for affective attachment in a rapidly modernizing world.
[33] Virilio, *Pure War*, 186.
[34] Ibid., 188, 189.
[35] Ibid., 186.
[36] Ibid., 203.
[37] Leerom Medovoi, "Global Society Must Be Defended: BIOPOLITICS WITHOUT BOUNDARIES," *Social Text* 25, no. 2 91 (Summer 2007): 66.
[38] Virilio, *Pure War*, 35.
[39] Wallace, *Infinite Jest*, 777.
[40] Medovoi, "Global Society Must Be Defended," 54.
[41] The creation and actions of O.N.A.N. reflect on an international scale Giorgio Agamben's theory of the "state of exception," which argues that the contemporary state exists in a permanent state of emergency in which the sovereign power of the state is extended beyond its usual limits in response to a real or perceived crisis. Wallace presents a state whose very existence is predicated on the state of exception: O.N.A.N. is created in response to a crisis situation and all of its actions are, according to its leaders at least, necessary responses to an ongoing emergency with no end in sight. Giorgio Agamben, *State of Exception* (Chicago: University of Chicago Press, 2005).
[42] Wallace, *Infinite Jest*, 571.
[43] Ibid., 183.
[44] Ibid., 573.
[45] Ibid. Emphasis in original.
[46] Ibid., 503.
[47] Nicholas Royle, "Nuclear Piece: Mémoires of Hamlet and the Time to Come," *diacritics: A Review of Contemporary Criticism* 20, no. 1 (1990): 45.

[48] Paul K. Saint-Amour, "Bombing and the Symptom: Traumatic Earliness and the Nuclear Uncanny," *diacritics: A Review of Contemporary Criticism* 30, no. 4 (2000): 65.
[49] Fredric Jameson, "New Literary History After the End of the New," *New Literary History: A Journal of Theory and Interpretation* 39, no. 3 (2008): 383.
[50] See Marshall Boswell, *Understanding David Foster Wallace* (Columbia: University of South Carolina Press, 2003) for a more complete version of the received wisdom on this.
[51] Frank Louis Cioffi, "'An Anguish Become Thing': Narrative as Performance in David Foster Wallace's *Infinite Jest*," *Narrative* 8, no. 2 (May 2000): 170.
[52] Daniel Carpenter Grausam, *On Endings: American Experimental Fiction and the Cold War* (Berkeley, CA: University of California Press, 2005), 192-3.
[53] For examples of the kinds of compulsions this novel produces, see The Internet.
[54] Wallace, *Infinite Jest*, 1011n.
[55] Jacques Derrida, *Of Grammatology*, Corrected ed. (Baltimore: Johns Hopkins University Press, 1998), 144-5.
[56] Derrida, Porter, and Lewis, "No Apocalypse, Not Now (Full Speed Ahead, Seven Missiles, Seven Missives)," 31.
[57] Samuel Beckett, *Molloy; Malone Dies; and, The Unnamable: Three Novels* (New York: Grove Press, 1959), 418.
[58] Virilio, *Pure War*, 203.
[59] Ibid., 223.
[60] Again, see The Internet; also Grausum, *On Endings*, 160-162.
[61] Royle, "Nuclear Piece," 54.
[62] Wallace, *Infinite Jest*, 559.
[63] Derrida, Porter, and Lewis, "No Apocalypse, Not Now (Full Speed Ahead, Seven Missiles, Seven Missives)," 31.
[64] Ibid.

Chapter Ten

The Hunger Games: Darwinism and Nuclear Apocalypse Narrative in the Post-9/11 World

Patrick B. Sharp

Jeffrey S. Kaplan observes that young adult literature has taken a turn toward darkness, violence, and dystopia in the aftermath of 9/11, and he cites Suzanne Collins' *Hunger Games* trilogy as a benchmark in the popularity of such literature.[1] *The Hunger Games* novels and the recent film owe much of their structure not to 9/11—and the subsequent wars in Afghanistan and Iraq—but rather to the apocalyptic and dystopian science fiction (SF) of the Cold War. The ecological apocalypse of *The Hunger Games*, and the threat of mutual assured destruction with nuclear weapons that haunts the sequels *Catching Fire* and *Mockingjay*, would have been right at home in the late 1950s and 1960s. Together they constitute yet another example of how Cold War narratives continue to circulate in the post 9/11 system of genres as people try to make sense out of the technologies that shape our present and threaten our future. At the same time, the popularity of the novels and the film show the continued importance of Darwinist visions of the frontier for American identity. While offering a critique of the faux-Darwinism of reality TV shows such as *Survivor*, Collins reinforces the belief that democracy and freedom grow from engaging in a struggle to survive on the frontier. In this regard, Collins' novels are actually extremely optimistic visions of the future in the mode of classic nuclear frontier fiction. While she contributes to this genre, however, Collins challenges the essentialist gender assumptions of Darwinist anthropology and the nuclear frontier stories of the Cold War.

Genre and Apocalypse

When discussing *The Hunger Games* trilogy, which was published between 2008 and 2010, it is important to attend to how the novels employ genre to create an immediately familiar milieu for their readers. Within science fiction studies, a recent paradigm shift has encouraged new ideas about genre that have changed our understanding of what science fiction is and how it emerged. In the 1960s and 1970s, such influential scholars as Darko Suvin and Samuel Delany produced definitions of science fiction that were based on formalist assumptions about genre. Their definitions provided the basis for measuring texts according to narrow, static criteria and excluding texts that did not measure up.[2] Their goals were taxonomic: they were trying to come up with definitions of a literary genre that could be used to sort texts into categories, and in the process to identify the essence of the genre. One major flaw with such formalist approaches to genre is that they try to pin down an essence that is constantly changing. Over the past few decades, scholars in the humanities and social sciences have developed accounts of genre that attempt to balance the analysis of formal textual properties with an understanding of how genres emerge and evolve.[3]

One way recent genre studies have avoided lapsing back into an unproductive taxonomic mode of analysis is to emphasize the "social and historical aspects" of any given text.[4] A genre, as defined by Tzvetan Todorov, is "the historically attested codification of discursive properties."[5] Genres comprise a system that is constantly shaping, and that is shaped by, the ideology and institutions of a particular culture.[6] Several recent theories of genre—in fields such as sociology, anthropology, linguistics, rhetoric, cultural studies, and film and television studies—have demonstrated that genres arise to solve recurring communicative or representational issues faced by members of a community.[7] However, genres do not spring fully formed from the minds of authors. As Todorov describes, genres come "from other genres. A new genre is always the transformation of an earlier one, or of several: by inversion, by displacement, by combination."[8] As a genre develops, certain formal elements are repeated and become codified because of their familiarity and success in handling communicative or representational issues.

One critical issue scholars have to address when engaging in genre analysis comes with the analysis of the genres active within a specific text. The system of genres that exist at the time of the text's production provide a complex set of discursive frames and formal structures that writers and readers have at their disposal when producing and interpreting texts. During the act of producing a text, its creators draw upon what M. M. Bakhtin calls both "primary" and "secondary" genres. Primary genres are simple and unmediated, whereas secondary genres are complex and generally constitute more formal discursive patterns such as novels, lab reports, and political speeches. During their development, secondary genres generally ingest or include primary genres.[9] With the presence of so many genres within any given text, it then becomes necessary to talk about a text in terms of its *complex* of genres, which draws from the numerous genres available in a culture.[10] Within a given culture, the available genres constitute a system that serves as a resource for authors to tap into when creating a text, and that readers rely upon when reading a text.

Scholarship on the origins of SF has long emphasized this aspect of genre: SF was cobbled together from pre-existing genres such as the gothic romance, the extraordinary voyage, the tale of the future, and the tale of science (otherwise known as the gadget story).[11] More recent scholarship in this vein, however, has emphasized *why* science fiction emerged when it did. This scholarship examines the social needs and historical conditions the genre addressed, and the entire system of genres that early science fiction authors drew upon when writing their stories. In his influential cultural history of the genre, Roger Luckhurst cites such developments as mass education in science and literature, the impact of industrialism on everyday life, and changes in the publishing industry as key "conditions of emergence" for science fiction.[12] John Rieder has argued that science fiction's "precursor" genres—and science fiction itself—"represent ideological ways of grasping the social consequences of colonialism."[13] As Rieder shows, and as I demonstrate in *Savage Perils*, science fiction was heavily influenced by such closely related colonial genres as the scientific race treatise, the travelogue, the "lost race" story, and new narratives of human evolution.

To understand *The Hunger Games* universe, therefore, it is necessary to understand the complex of genres that the texts draw upon that help make them so popular. One of the major genres Collins uses in her novels is the ecological apocalypse. The popularity of ecological apocalypse grew from the nuclear fear of the late 1950s. That decade saw an exponential growth in nuclear technologies from Nagasaki-like implosion devices to boosted thermonuclear fusion bombs that created huge lethal clouds of fallout.[14] A generation had been exposed to various levels of propaganda and military-style drills through civil defense programs that emphasized the dangers of nuclear weapons.[15] The arms race between the United States and Soviet Union—and the above-ground testing that was a central part of early Cold War weapons development—fanned fears about the toxicity of nuclear technologies. Members of the growing anti-nuclear movements questioned what damage was being done to the environment through such weapons testing. Rachel Carson published her hugely influential book *Silent Spring* in late September of 1962, just a couple of weeks before the Cuban Missile Crisis raised the nuclear fear of Americans to an all-time high. As Ralph H. Lutts shows, Carson capitalized on the familiarity of her audience with radioactive fallout to make her case about the dangers of pesticides such as DDT.[16]

In the opening chapter entitled "A Fable for Tomorrow," Carson translates the imagery of the nuclear wasteland and apocalyptic SF into the imagery of environmental devastation that she says has actually occurred in "real communities" throughout the country.[17] One common focus of SF stories in the early Cold War was the disruption nuclear technologies brought to reproduction, family life, and the domestic sphere. Stories such as Judith Merril's "That Only a Mother" (1948) dramatized the dangers that nuclear technologies posed to unborn life, and warned how nuclear technologies could mutate the future that is embodied in children. Carson hits this point twice in the opening chapter, noting that in her imagined future community, "There had been several sudden and unexplained deaths, not only among adults but even among children, who would be stricken suddenly while at play and die within a few hours... On the farm the hens brooded, but no chicks hatched. The farmers complained that they were unable to raise any pigs—the litters were small and the young survived only a few days."[18] Here Carson employs the poignant image of children at play to emphasize the innocence that environmental contamination destroys. The failure of livestock to reproduce shows the danger to the food supply, bringing into focus the possibility that survival might not be possible for anyone if the poisoning continues. Carson makes the connection between chemical pesticides and nuclear fallout explicit with the image of a mysterious "white granular powder" that had "fallen like snow upon the roofs and the lawns, the fields and streams" and collected "in the gutters under the eaves and between shingles of the roofs."[19] Thus Carson's fable paints a picture of indiscriminant death that threatens the domestic sphere, the family, and reproduction itself.

In case the connection between nuclear fallout and pesticides is still unclear, Carson states her case more plainly early in the second chapter. She argues that,

> In this now universal contamination of the environment, chemicals are the sinister and little-recognized partners of radiation in changing the very nature of the world—the very nature of its life. Strontium 90, released through nuclear explosions into the air, comes to earth in rain or drifts down as fallout, lodges in soil, enters into the grass or corn or wheat grown there, and in time takes up its abode in the bones of a human being, there to remain until his death. Similarly, chemicals sprayed on croplands or forests

or gardens lie long in soil, entering into living organisms, passing from one to another in a chain of poisoning and death.[20]

Carson was a trained biologist, and for her evolution was at the center of the problem brought on by these artificial technologies of death. Explaining the evolutionary dangers of new chemicals, she argues that, "The rapidity of change and the speed with which new situations are created follow the impetuous and heedless pace of man rather than the deliberate pace of nature... To adjust to these chemicals would require time on the scale that is nature's; it would require not merely the years of a man's life but the life of generations."[21] In other words, Carson believes that the pace of modern technological progress has outstripped the pace of evolutionary change; she believes that we need to create lives and technologies more in harmony with the pace of nature, or risk apocalypse.

Evolution and The Apocalyptic Vision of *The Hunger Games*

Carson's image of apocalypse in *Silent Spring* was very popular and had a profound influence on generations of environmentalists. In the years since its first publication, images of ecological disaster have often gone hand in hand with future visions of nuclear disaster. William Gibson's *Neuromancer* (1984) and Marge Piercy's *He, She, and It* (1991) are just two well-known examples of this phenomenon, where limited nuclear wars have contributed to the accelerating degradation of the environment and the collapse of existing political structures. In post-9/11 American culture, the threat of nuclear weapons has been attached to the threat of terrorism in some obvious ways in highly rated television shows such as *24* and *NCIS*. The 2003 reboot of *Battlestar Galactica* revived the Cold War narrative of what Senator John F. Kennedy called a "nuclear Pearl Harbor," a fear that was at the core of the original 1978 series. These shows make clear that nuclear threats are not always linked to the larger issue of environmental contamination, and that "nuclear terrorist" and "nuclear Pearl Harbor" stories still have a great deal of currency in our contemporary system of genres.

With that said, *The Hunger Games* universe is more in the vein of Rachel Carson's *Silent Spring*: the nuclear issues of the novels and the film are subsumed within a larger struggle to survive in the face of environmental devastation. This is made clear early in the first novel with the preparations for the "reaping" ceremony. The reaping is where young people—who are the age of the book's target audience—are lined up and two of their names are chosen in a lottery to participate in the Hunger Games. Katniss Everdeen is a teenager who is the central protagonist and narrator of the novels. As part of the ceremony, we are given through her perspective a condensed "history of Panem," a propaganda piece that explains how "the country... rose up out of the ashes of a place that was once called North America."[22] The Mayor of the area called District 12 "lists the disasters, the droughts, the storms, the fires, the encroaching seas that swallowed up so much of the land, the brutal war for what little sustenance remained."[23] Collins creates a hardened and jaded voice for Katniss that is painfully familiar with this apocalyptic story. In the 2012 film based on the first novel, this information is given in the form of a propaganda reel that includes a nuclear mushroom cloud, the dominant Cold War image associated with apocalypse. Thus the series of disasters brought about by climate change in Collins' novel actually precipitates a war where nuclear weapons play a significant role. The ecological disasters trigger a nuclear disaster and the ongoing threat of extinction via nuclear weapons.

In representing life after an apocalypse, Collins draws on the American ideal of the frontier and the nuclear frontier stories of the Cold War. Katniss displays many characteristics associated with the frontiersman, the heroic American figure that has been central to the national identity of the United States since at least the eighteenth century. Historical figures such as Daniel Boone were celebrated as masters of both civilized politics and wild nature after the Revolutionary War, and were used to establish a unique identity for the United States that was separate from Great Britain. The image of the frontiersman popularized by authors such as John Filson and James Fenimore Cooper embodied the rugged origins of American democracy: the heroism and virtue of frontiersmen marked them as superior to the "savages" of the Americas and the decaying civilizations of Europe.[24] In the late nineteenth century, historian Frederick Jackson Turner reimagined the frontiersman as a heroic individual who had evolved in the struggle to survive on the rugged American continent, and made the frontiersman central to his narrative of United States history.[25] Turner argued that, "the most important effect of the frontier has been in the promotion of democracy here and in Europe. As has been indicated, the frontier is productive of individualism… The frontier individualism has from the beginning promoted democracy."[26] Turner's Darwinist vision of the frontier past became projected into the apocalyptic futures of nuclear frontier stories that emerged in the 1950s. In the ongoing Cold War struggle between the United States and the Soviet Union, frontier imagery was used to both criticize the dangers of nuclear proliferation and ensure Americans that they would survive a nuclear war.[27]

Collins' use of the nuclear frontier genre becomes apparent early in the story. The struggle to survive faced by Katniss and her family after her father's death in a mining explosion forces her to take up a bow and arrow and go hunting in the forests near her prison camp of a home in the Appalachian Mountains. Before the reaping, Katniss climbs through the electrified fence surrounding her coal-mining town and gets her "bow and sheath of arrows from a hollow log."[28] Katniss notes that, "more people would risk it if they had weapons. But most are not bold enough to venture out with just a knife. My bow is a rarity, crafted by my father along with a few others that I keep well hidden in the woods... My father could have made good money selling them, but if the officials found out he would have been publicly executed for inciting a rebellion."[29] Here Collins establishes Katniss' bravery and individualism, two hallmarks of the frontiersman. However, Collins goes far beyond that: by simply having her bow and arrows, Katniss is marked as rebelling against a totalitarian regime. Katniss is represented as having a unique mastery over wild nature, and her struggle to survive in the forest leads to a keen sense of independence. When thrust onto a national stage in the Hunger Games, Katniss cannot help but defy the authority of the central government in a manner that fuels a new democracy movement. In essence, Collins repeats Turner's story of American independence through the character Katniss.

Collins clarifies the advantages Katniss has developed because of her frontier life after the reaping. Katniss' delicate younger sister Prim is chosen in the lottery, and Katniss volunteers to replace her. Katniss is immediately shipped off to the Capitol to prepare for the televised games where she will battle twenty-three other teenagers to the death. When Katniss has her first training session and sees her fellow "tributes" for the first time, she observes that,

> Almost all of the boys and at least half of the girls are bigger than I am, even though many of the tributes have never been fed properly. You can see it in their bones, their skin, the hollow look in their eyes. I may be smaller naturally, but overall my family's resourcefulness has given me an edge in that area. I stand straight, and while I'm thin, I'm strong. The meat and plants from the woods combined with the exertion it took to get them have given me a healthier body than most of those I see around me.[30]

Collins portrays Katniss' real struggle to survive and provide for her family as something that has made her fitter than the competition. The "career tributes" that come from more privileged districts have never missed a meal, and because of this they are represented as coddled and weak. This is another classic nuclear frontier trope: the heroes' frontier survival skills distinguish her from those who are weaker due to their relationship with a system that isolates them from a true Darwinian struggle for survival. The fact that the Capitol citizens and the career tributes have never truly struggled to survive makes them objects of scorn; they are over-civilized in ways that ultimately lead to their defeat. In this way, Katniss' life is a romanticized vision of American origins: the ecological and nuclear apocalypses have simply returned her to this "lost," primal, Darwinian mode of existence on the nuclear frontier.

Gender and the Nuclear Frontier

One major difference between *The Hunger Games* universe and the nuclear frontier stories of the early Cold War comes with Collins' representation of gender. The Darwinism of Turner and countless others always assumed that the evolutionary hero of the frontier was a male. This can be traced back directly to Charles Darwin's *The Descent of Man* (1871), where he expanded upon the relevance of his theory of "natural selection" for understanding human origins. A major portion of this work was devoted to "sexual selection," a mechanism that Darwin saw as counterbalancing natural selection. In Darwin's representation of human development, sex was directly related to technology: he believed that men were by nature violent, selfish toolmakers that passed these traits on to their sons.[31] Darwin linked the traits he associated with human progress such as upright posture, bigger brains, and the development of tools to males. Darwin mused that men have "to defend their females, as well as their young, from enemies of all kinds, and to hunt for their joint subsistence."[32] He went on to assert that males have been selected for "energy, perseverance, and courage" as well as "higher powers of imagination and reason."[33] Darwin reasoned that these traits "will have been developed in man, partly through sexual selection,—that is, through the contest of rival males, and partly through natural selection."[34] Like generations of male scientists before him, Darwin defined women in terms of motherhood. Because of their biology, Darwin asserted that women were nurturing, beautiful prizes to be fought over by the men for reproductive purposes. This meant that men were natural scientists, soldiers, and hunters, not women, and the only important contribution women made to evolution was through reproduction and their choice of mates.[35] In the end, Darwin states that "man has ultimately become superior to woman" through the processes of natural and sexual selection.[36] This vision of evolution was so persistent in the twentieth century that Ruth Hubbard posed the rhetorical question, "Have only men evolved?"[37]

In Turner's vision of frontier evolution, he accepted Darwin's formulation of sexual selection and assumed that it was males who drove progress and the development of American democracy. This male frontier hero served as the central figure in countless nuclear frontier stories during the early Cold War. In the 1960s, Darwin's account of human evolution became developed into the narrative labeled "man the hunter." Hunting, it was argued, caused the rise of the nuclear family: men had to sort out which women were theirs so that they wouldn't fight while out on a hunt. The meat brought in by hunting was claimed to have fueled the development of brains both in size and cognitive abilities.[38] In the 1970s, anthropologists such as Sally Slocum argued that "woman the gatherer" had actually formed the core of the nuclear family, and the cognitive demands of foraging had fueled brain development. Slocum preserved an essentialist formulation of gender that had been in place since Darwin's *Descent of Man*, with women staying near home and men going out into the hunting workplace.[39] This basic gendered division of labor was evident in science fiction of the 1960s in films such as *2001: A Space Odyssey*,[40] and it still constitutes a common essentialist formulation in evolutionary arguments today.

Collins challenges the essentialist formulation of the evolutionary hero, and the nuclear frontier hero, as a male. In this regard, she continues the feminist critique of normative gender roles in SF that has been common in the genre since at least the 1970s. While Gale, one of Katniss' two love interests, is represented as a skilled hunter, it is Katniss who is represented as superior with a bow and arrow. Gale even goes so far to tell Katniss that she is likely to do well in the Hunger Games because, "Katniss, it's just hunting. You're the best hunter I know."[41] Her other love interest Peeta points out that, "She's excellent... my father buys her squirrels. He always comments on how the arrows never pierce the body. She hits every one in the eye. It's the same with the rabbits she sells the butcher."[42] However, Collins' emphasis on Katniss' hunting is not some simple kind of gender inversion: Katniss is also represented at several points as a superior gatherer as well. The fact that she learned both hunting and gathering from her father further undermines the notion that such survival skills are somehow linked to gender. Like her father, Katniss is a combination of man the hunter and woman the gatherer, a heroic frontierswoman with diverse survival skills that can be learned from, and taught to, both sexes.

In addition to undermining the Darwinist gendered assumptions built in to survival skills, Collins also shows the performative and contingent nature of romance. Early on, Collins shows how disinterested Katniss is in the trope of evolutionary romance central to Darwin's theory of sexual selection. When evaluating romance at the beginning of the novel, Katniss describes Gale in terms of the ideal Darwinian male: "Gale won't have any trouble finding a wife. He's good-looking, he's strong enough to handle the work in the mines and he can hunt."[43] Gale is seen as a strong man and provider who can easily attract a mate. However, because of the danger posed to children by the Hunger Games and her difficult life, Katniss makes clear that she doesn't want a romantic partner. In order to gain the sympathy of the audience, and to get special gifts during the televised Hunger Games, Katniss is later forced to engage in a romantic narrative that is completely out of character. When Peeta reveals to the television audience (and to Katniss) that he loves her, Katniss grudgingly accepts this only because it is a survival strategy. Though Katniss warms to Peeta, eventually sharing "the first kiss that makes [her] want another," Katniss still focuses primarily on survival.[44] In one sense, Katniss' attraction to Peeta can be read as confirming Darwin's concept of sexual selection. We learn that after the death of Katniss' father, Katniss was probably going to starve to death when Peeta braved pain and violence in order to give her a loaf of bread. When mulling over her feelings for Peeta, Katniss realizes that "I have kept track of the boy with the bread."[45] Thus Collins shows Katniss' attraction to Peeta as rooted in his providing for her at a key moment of her life. Because he saves her, she appears to have a deep and unconscious attraction to him.

On the other hand, Collins' representation of their survival skills undermines any simplistic reading of the Katniss/Peeta relationship as a Darwinist romance. When the popularity of the staged romance proves popular with the audience, the "Gamesmakers" who control the arena decide they will allow two winners for the first time. Because both winners would have to be from the same district, Katniss goes in search of Peeta in order to team up and improve her chances of survival. However, once they are together and moving through the forest in search of food, Katniss discovers that, "even on the smooth bed of needles, Peeta is loud. And I mean *loud* loud, as if he's stomping his feet or something... 'You've got to move more quietly,' I say. 'Forget about Cato, you're chasing off every rabbit in a ten-mile radius.'"[46] Where Gale had a "velvet tread" that made him a perfect hunting partner back home, Peeta is a liability both in regard to finding food and fending off the other tributes in the arena such as Cato. Peeta is a sensitive artist, and yet he ultimately wins out over the seemingly perfect Darwinian male Gale in the contest for Katniss' affections. In this way, Collins calls into question the inevitability of any essential romantic response based on the terms set out by Darwin.

Collins also undermines the notion that characters embodying aspects of Darwinist femininity are weak. Katniss' mother is introduced in the first novel as a hysterical woman whose emotional collapse after the death of her husband forced Katniss into the role of family provider. Later in the novel, Katniss has to confront the unique strength of her mother and her delicate younger sister, Prim, who work in their village as healers. The role of healer is consistent with Darwin's assertion that women are inherently nurturing due to their "maternal instincts."[47] When trying to nurse Peeta back to health during the games, Katniss has to confront his wound that is, "a deep inflamed gash oozing both blood and pus."[48] Katniss' response is to want to "run away... Go out and hunt while my mother and Prim attend to what I have neither the skill nor the courage to face."[49] Katniss eventually has to find the strength to nurse Peeta because she is the only one who can save him. Collins uses Katniss' change of perspective to show a different type of frontier strength and survival that goes far beyond hunting and gathering. This feminine form of nursing is celebrated as requiring a bravery and strength of its own. Collins shows Katniss eventually embracing this role, in effect taking it on in the name of survival instead of some feminine essence. Collins shows Katniss as a victor, a hero capable of taking on the survival skills associated with both men and women in a way that makes her superior to her competitors (including Peeta). Collins therefore thwarts any simple gender stereotypes and calls into question both the essential nature and value attributed to gender roles by Darwin.

Mutually Assured Destruction and Frontier Democracy

In *Catching Fire* and *Mockingjay*, Collins highlights how Katniss' frontier struggle to survive has fueled her independence, a key aspect of her superior fitness, and this is the catalyst needed to return Americans to the democratic government favored by her ancestors. In one sense, this is a nostalgic and utopian vision, where young people are able to fix the problems created by their parents and return to an idealized state. In another sense, this is a hopeful vision of how we can actually evolve in a way that allows us to overcome the ecological apocalypse that Carson worried might overwhelm our capacity to adapt. In *Catching Fire*, the President of Panem is furious at Katniss because of her rebellious actions in the first novel. With open rebellions popping up all over, President Snow orchestrates a special seventy-fifth anniversary edition of the Hunger Games where two previous winners from each of the districts will battle to the death. As the only female victor of the games from District 12, Katniss is assured of a place in the games. With Peeta once again at her side, Katniss thinks that she will try and orchestrate the games so that Peeta will survive and return home. What Katniss doesn't know (and what the readers only suspect) is that many of the other former winners have agreed to sacrifice themselves to save Katniss so that she can lead the rebellion. When attacked by a lethal fog of nerve gas in the arena, the allied members of Katniss' group try to flee down a hill. Katniss is trying to rescue an elderly tribute named Mags by carrying her, but when Katniss stumbles, "Mags hauls herself up, plants a kiss on Finnick's lips, and then hobbles straight into the fog."[50] Katniss witnesses this sacrifice, and then recounts how, "Some deep-rooted animal desire for survival keeps me stumbling after Finnick and Peeta, continuing to move, although I'm probably dead already."[51] Aided by the sacrifice of other tributes and driven by her own inescapable "desire for survival," Katniss emerges once again as a frontier hero worthy of leadership. Collins shows Katniss as a flawed survivor, someone who is the perfect leader precisely because she is not a polished, over-civilized monster like President Snow.

In *Mockingjay*, Collins turns to another realm of Cold War nuclear narrative. With Katniss rescued from the arena by rebels, she is taken to the mysterious District 13 to serve as a figurehead for the rebellion. Collins characterizes District 13 as originally designed to be,

> a clandestine refuge for government leaders in time of war or a last resort for humanity if life above became unlivable. Most important for the people of 13, it was the center of the Capitol's nuclear weapons development program. During the Dark Days, the rebels in 13 wrested control from the government forces, trained their nuclear missiles on the Capitol, and then struck a bargain: They would play dead in exchange for being left alone. The Capitol had another nuclear arsenal out west, but it couldn't attack 13 without certain retaliation. It was forced to accept 13's deal.[52]

Here Collins evokes the stalemate of Mutually Assured Destruction (MAD), that Cold War strategy characteristic of both Soviet and American military planning. District 13 is basically a vast nuclear bunker, its extensive vaults and regimented civil defense planning a dystopian vision of centralized control. Caught between two totalitarian powers, Katniss becomes acutely aware of her own role as a pawn. Though she supports the rebellion, her frontier sense of independence gives her an immediate and pervasive dislike for District 13 and its leader, a grim woman named President Coin. Where Panem's President Snow represents the excesses of colonial capitalism, Collins uses President Coin to show the dangers of rigid, humorless, and lifeless socialism.

Collins dramatizes Katniss' rebellion against both totalitarian regimes when she is sent to view a recently bombed District 8 that has been punished by the Capitol. Katniss is working with a former Gamemaker-turned-rebel named Plutarch, the man who is now in charge of propaganda for District 13 and the rebellion. When asked to pose and read canned lines, Katniss comes across as horrible and stiff. However, when the Capitol launches another surprise attack on District 8, Katniss ignores direct orders and springs into action. Armed with a new high-tech bow and with Gale now at her side, Katniss charges up a building and observes "Seven small bombers in a V formation. 'Geese!' I yell at Gale. He'll know exactly what I mean. During migration season, when we hunt fowl, we've developed a system of dividing the birds so we don't both target the same ones... I estimate the lead time on the hoverplanes and let my arrow fly. I catch the inside wing of one, causing it to burst into flames."[53] Again, Collins evokes Katniss' role as the frontier hero to make her a true leader of the war against tyranny. Katniss' frontier survival skills—in this case, the hunting of geese—and her independent streak turn her into a true war hero. When painted up for the cameras, Collins represents Katniss as fake and ineffective. However, when she is engaged in a struggle to survive like on the frontier, Katniss is represented as "real," and therefore a more effective leader for the rebellion. This shows the value Collins places on this sense of traditional American individualism embodied in the Darwinist hero.

As the war unfolds, Collins again blends the threat of ecological apocalypse and nuclear apocalypse. When District 13 is attacked by missiles, the rebels doubt that it is a nuclear attack because "[n]uclear missiles would release radiation in the atmosphere, with incalculable environmental results."[54] Nuclear weapons are a threat here not because of their blast effects, but because of their lingering toxicity. The toxic threat of nuclear weapons is compounded by the simple threat of insufficient numbers needed for the species to survive. Boggs, a high-ranking officer in District 13 who is helping run the rebellion, says, "But there's always the larger question: If we engage in that type of war with Capitol, would there be any human life left?... You'll notice neither side has launched nuclear weapons. We're working it out the old-fashioned way."[55] Collins' text is decidedly optimistic in this conceit that human beings will make a rational calculation against using nuclear weapons. In fact, Collins goes so far as to make even conventional warfare seem like an unsurvivable prospect. As the rebels get closer to the Capitol, their tactics have to change because their leading scientist worries that if they continue to simply kill the opposition, they might be leaving such a shallow gene pool that humans might be "killing ourselves off."[56]

Collins takes a few more utopian turns in establishing the hope for a democratic republic after the apocalypse. Most of the rebels from outside of District 13 share Plutarch's hope for how the war will end: "'We're going to form a republic where the people of each district and the Capitol can elect their own representatives to be their voice in a centralized government. Don't look so suspicious; it's worked before.'"[57] Katniss gives voice to the skepticism of the younger generation when she responds, "Frankly, our ancestors don't seem much to brag about. I mean, look at the state they left us in, with the wars and the broken planet."[58] Here Collins evokes the hopeful premise for her young adult audience: it is indeed possible for young people to correct the mistakes of their elders. As the novel progresses, the moral bankruptcy of the older generation becomes epitomized by President Snow and President Coin. President Snow is a ruthless dictator who uses the Hunger Games to kill children and keep the districts in check. In the final battle of the war, Katniss learns how low the older generation will go for power. As the rebels approach President Snow's mansion, Katniss witnesses her young sister Prim charging in as a medic to aid some injured civilians. Then a hovership with Capitol markings bombs the civilians, including Prim, committing the final atrocity that turns the few remaining Capitol supporters against President Snow. The trauma of the Hunger Games, the subsequent war, the blast from this explosion, and the loss of her beloved sister leave Katniss feeling that "inwardly, I'm such a wasteland."[59] This trauma serves as another dimension of Collins' statement against warfare, nuclear weapons, and environmental devastation. The apocalypses of her forebears and the war of the older generation have caused Katniss to internalize the landscape of the wasteland. She is left as a shell of herself, suffering from Post-Traumatic Stress Disorder and unable even to think about the lingering romantic triangle with Gale and Peeta. In this regard, Collins' work provides a stark contrast to other young adult literature such as the *Twilight* universe of Stephenie Meyer.

However, Collins' hope for human evolution is still borne out by Katniss and her frontier spirit. When the Capitol is defeated and President Snow is imprisoned, Katniss confronts President Snow about his supposed murder of Prim and the civilians. This is where she pieces together all of the information and uneasiness she has felt toward President Coin.

President Snow quite logically breaks it down for Katniss: "Well, you really didn't think I gave the order, did you? Forget the obvious fact that if I'd had a working hovercraft at my disposal, I'd have been using it to make an escape... My mistake... was being so slow to grasp Coin's plan. To let the Capitol and the districts destroy one another, and then step in to take power with Thirteen barely scratched."[60] Katniss remembers that this civilian bombing fits the profile of a trap Gale had developed and given to the leaders of the rebellion. In the end, the top leaders of both sides are corrupt. As the figurehead of the rebellion, however, Katniss is given the honor of executing President Snow with her iconic bow and arrow in front of the nation. She instead aims slightly to the side and kills President Coin, with President Snow expiring from his own lingering illness. With both potential dictators eliminated by Katniss and her frontier heroism, Collins makes her final turn toward a utopian future. With the establishment of the republic, Plutarch muses to Katniss, "We're fickle, stupid beings with poor memories and a great gift for self-destruction. Although who knows? Maybe this time... it sticks. Maybe we are witnessing the evolution of the human race."[61] This hopeful ending culminates in Katniss picking a man. Gale, the ideal Darwinian love interest, has become corrupted by his penchant for violence and his collusion with the older generation. Peeta, the sensitive artist, never cooperated. The novel ends with Peeta and Katniss helping each other through the trauma of their childhood and building a better future for their own children, a future built on the dark lessons of the past. Instead of following some sort of overwhelming, primal drive to mate with a superior provider and fighter, Katniss makes a choice guided by morality and her hope for a better future. While this preserves the heteronormative romance plot, Collins avoids falling back into the simple essentialist logic of sexual selection.

While it is essential for Nuclear Criticism to attend to the prospect of our survival as a species, it is equally as important for Nuclear Criticism to interrogate the meanings we attribute to survival. Charles Darwin forever changed Euro-American concepts of survival, and this has been reflected in the Darwinist narratives we have told each other for the past one hundred and fifty-three years. Darwin's work on evolution arose within the context of European colonial expansion, and his evolutionary narratives bear the indelible stamp of their origins. Darwin's account of human evolution proved popular in part because it spoke to Victorian beliefs about colonization, race, and gender. While many of his insights have stood the test of time, Darwin's arguments about colonization, race, and gender have been subject to extensive critique and revision. Though we have left many Victorian beliefs behind, Darwinist formulations of colonization, race, and gender are still encoded within many of the narratives we tell ourselves, especially those dealing with survival. Nuclear Criticism must always attend to the colonial Victorian baggage that comes with representations of survival in the face of nuclear technologies. Nuclear Criticism must also come to grips with how nuclear technologies have become subsumed under larger ecological visions of apocalypse and survival.

Collins' *Hunger Games* universe provides an interesting case study in how nuclear narratives have been folded into ecological narratives. In this sense, *The Hunger Games* adds an interesting new chapter to the nuclear frontier genre of science fiction. Even though the genre came of age in the Cold War, it clearly still resonates with readers too young to know about "Duck and Cover" or even the terrorist attacks of 9/11. These nuclear frontier narratives are deeply entrenched in the national psyche of the United States, and still have a great deal of power to help us make sense of the present by imagining the future. Collins' novels are ultimately utopian in a number of ways: they posit that teenagers can overcome ecological catastrophes, limited nuclear wars, and the threat of mutually assured destruction in order to build a democracy that is sustainable. They also imagine a world where young women can learn to appreciate and embody many different types of strength and master survival skills without being limited by normative notions of gender. In the end, Collins provides the hopeful thought that we might be able to evolve quickly enough to avoid extinction via nuclear and ecological apocalypse. Her work shows that nuclear weapons have not been forgotten. Indeed, Collins' work provides a clear example of how young people are still learning about the dangers of nuclear technologies. By foregrounding the toxicity of nuclear weapons, *The Hunger Games* trilogy continues the tradition of Rachel Carson in contextualizing the dangers that nuclear technologies pose to all life.

Notes

[1] Jeffrey S. Kaplan, "The Changing Face of Young Adult Literature: What Teachers and Researchers Need to Know to Enhance Their Practice and Inquiry," in *Teaching Young Adult Literature Today: Insights, Considerations, and Perspectives for the Classroom Teacher*, eds. Judith A. Hayn and Jeffrey S. Kaplan (Lanham, MD: Rowman & Littlefield, 2012), 24-25.
[2] Roger Luckhurst, *Science Fiction* (Malden, MA: Polity, 2005), 6-7.
[3] For an overview of genre theory in many different disciplines, see Charles Bazerman, "Social Forms as Habitats for Action," *Journal of The Interdisciplinary Crossroads* 1 (2003): 123–42. For the current state of genre theory in relation to

science fiction, see John Rieder, "On Defining SF, or Not: Genre Theory, SF, and History," *Science Fiction Studies* 37, no. 2 (2010): 191-209.

[4] Carolyn R. Miller, "Genre as Social Action," in *Genre and the New Rhetoric*, eds. Aviva Freedman and Peter Medway (Bristol, PA: Taylor and Francis, 1994), 24.

[5] Tzvetan Todorov, *Genres in Discourse,* trans. Catherine Porter (New York: Cambridge University Press, 1990), 9.

[6] Ibid., 18-19.

[7] Bazerman, "Social Forms;" Pierre Bourdieu, *Language and Symbolic Power*, trans. by Gino Raymond and Matthew Adamson, ed. John B. Thompson (Cambridge, MA: Harvard University Press, 1994), 129; Thomas Luckmann, "On the Communicative Adjustment of Perspectives, Dialogue and Communicative Genres," in *The Dialogue Alternative*, ed. Astri Heen Wold (Oslo: Scandinavian University Press, 1992), 228.

[8] Todorov, *Genres in Discourse*, 15.

[9] M. M. Bakhtin, *The Dialogic Imagination,* trans. by Caryl Emerson and Michael Holquist, ed. by Michael Holquist (Austin: University of Texas Press, 1981), 60-63.

[10] Todorov, *Genres in Discourse*, 10.

[11] Edward James, *Science Fiction in the Twentieth Century* (New York: Oxford University Press, 1994), 13-26; Paul Kincaid, "On the Origins of Genre," *Extrapolation* 44, no. 4 (2003): 423.

[12] Luckhurst, *Science Fiction*, 16-29.

[13] John Rieder, *Colonialism and the Emergence of Science Fiction* (Middletown, CN: Wesleyan University Press, 2008), 20.

[14] Mark Hamilton Lytle, *The Gentle Subversive: Rachel Carson,* Silent Spring*, and the Rise of the Environmental Movement* (New York: Oxford University Press, 2007), 142-44; Spencer Weart, *Nuclear Fear: A History* (Cambridge, MA: Harvard University Press, 1988), 296-299.

[15] Andrew D. Grossman, *Neither Dead Nor Red: Civilian Defense and American Political Development During the Early Cold War* (New York: Routledge, 2001), 45-67; Laura McEnaney, *Civil Defense Begins at Home: Militarization and Everyday Life in the Fifties* (Princeton: Princeton University Press, 2000), 68-78; Patrick B. Sharp, *Savage Perils: Racial Frontiers and Nuclear Apocalypse in American Culture* (Norman: University of Oklahoma Press, 2007), 170-194.

[16] Ralph H. Lutts, "Chemical Fallout: Rachel Carson's *Silent Spring*, Radioactive Fallout, and the Environmental Movement," *Environmental Review* 9, no. 3 (1985): 212-18.

[17] Rachel Carson, *Silent Spring* (New York: Houghton Mifflin, 2002), 3.

[18] Ibid., 2.

[19] Ibid., 3.

[20] Ibid., 6.

[21] Ibid., 7.

[22] Suzanne Collins, *The Hunger Games* (New York: Scholastic, 2008), 18.

[23] Ibid.

[24] Sharp, *Savage Perils*, 16-20; Richard Slotkin, *Regeneration Through Violence:*

The Mythology of the American Frontier, 1600-1860 (New York: HarperPerennial, 1996), 268-276.
[25] Sharp, *Savage Perils*, 58-63.
[26] Frederick Jackson Turner, *The Frontier in American History* (New York: Dover, 1996), 30.
[27] Sharp, *Savage Perils*, 170-94.
[28] Collins, *Hunger Games*, 5.
[29] Ibid.
[30] Ibid., 94.
[31] Penelope Deutscher, "The Descent of Man and the Evolution of Woman," *Hypatia* 19, no. 2 (2004): 44.
[32] Charles Darwin, *The Descent of Man; and Selection in Relation to Sex* 2nd ed. (Amherst, New York: Prometheus, 1998), 584.
[33] Ibid., 585.
[34] Ibid.
[35] Ibid., 583-585; Ruth Hubbard, "Have Only Men Evolved?" in *The Gender of Science*, ed. Janet A. Kourany (Upper Saddle River, NJ: Prentice, 2002), 158-161; Joan Roughgarden, *Evolution's Rainbow: Diversity, Gender, and Sexuality in Nature and People* (Berkeley, CA: University of California Press, 2004, 2009), 164-172.
[36] Darwin, *The Descent of Man*, 585.
[37] Hubbard, "Have Only Men Evolved?"
[38] Sally Slocum, "Woman the Gatherer: Male Bias in Anthropology," in *Toward an Anthropology of Women*, ed. Rayna Rapp Reiter (New York: Monthly Review Press, 1975), 38-42.
[39] Ibid., 43-48.
[40] Robert Savage, "Paleoanthropology of the Future: The Prehistory of Posthumanity in Arthur C. Clarke's *2001: A Space Odyssey*," *Extrapolation* 51, no. 1 (2010): 99-112.
[41] Collins, *Hunger Games*, 40.
[42] Ibid., 89.
[43] Ibid., 10.
[44] Ibid., 298.
[45] Ibid., 93.
[46] Ibid., 314.
[47] Darwin, *The Descent of Man*, 583.
[48] Collins, *Hunger Games*, 256.
[49] Ibid.
[50] Suzanne Collins, *Catching Fire* (New York: Scholastic, 2009), 301.
[51] Ibid., 302.
[52] Suzanne Collins, *Mockingjay* (New York: Scholastic, 2010), 17.
[53] Ibid., 96.
[54] Collins, *Mockingjay*, 138.
[55] Ibid., 81.
[56] Ibid., 206.

[57] Ibid., 83-84.
[58] Ibid., 84.
[59] Ibid., 366.
[60] Ibid., 356-57.
[61] Ibid., 379.

CHAPTER ELEVEN

LEGACY WASTE:
NUCLEAR CULTURE AFTER THE COLD WAR

DANIEL CORDLE

> You are now in the tunnel. This place is not a place of honor. No esteemed deeds are commemorated here. You should not have come here. You are heading towards a place where you should never go. What is there is dangerous and repulsive. The danger will still be present in your time, as it is in ours. Please turn around and never come back. There is nothing here for you. Go no further.
> —Michael Madsen, *Into Eternity* (Denmark, 2010)

Into Eternity, a documentary by the Danish director, Michael Madsen, tells the story of Onkalo (literally: "hiding place") in Finland, where the world's only permanent storage facility for nuclear waste is being carved out of the rock. Designed to contain radioactive materials until they are no longer dangerous, Onkalo must last, we are told, for 100,000 years.

It could be—it is designed to be—the most enduring work of our civilization, but although the engineering feat is remarkable it is the compelling cultural and philosophical issues raised by Onkalo that are the focus of the film. The ambition to create an architecture that lasts for 1000 centuries containing not our legacies, but our legacy waste, raises intriguing questions.[1] Should this architecture be marked with warnings of the dangerous materials it contains, or when it is finally complete (Madsen notes that construction began when he was a child and will not be complete until the twenty-second century, long after his death) should all traces be buried in the hope that no-one finds it? If the former, who will forbid access to Onkalo in 100, 500 or 1000 years, let alone 10,000 or 50,000? What knowledge of nuclear toxicity will future societies have? What languages will people speak and how can we warn them not to excavate Onkalo's tunnels?

The most striking sequences of the film, in which Madsen speaks to camera or in voiceover, suppose a far future ingression at Onkalo.

Addressing himself to people imagined to have found the facility—archaeologists or treasure hunters, perhaps; the curious; the unlucky—Madsen constructs Onkalo as a place of ominous legend. As the film goes on, Onkalo is imagined to be penetrated deeper and deeper by these future explorers until, eventually, they reach the dangerous heart of the facility:

> You have now gone deeper into the tunnel and you have reached a place where you should never have come. Down here radiation is everywhere. You do not know it, but something is happening to your body right now. It is beyond your senses. You feel nothing. You smell nothing. An invisible light is shining right through you. It is the last glow of my civilisation, that harvested the powers of the universe.[2]

This vertiginous encounter with, if not quite the eternity of the film's title, certainly a period of time that dwarfs human culture thus far, defamiliarizes us productively from conceptions of time anchored to familiar reference points of human lifespans and recorded history. Our civilization's most abiding legacies might surprise us: not, perhaps, the spectacular, hubristic sky-scraping edifices currently springing up around the globe, but a hidden, tomb-like architecture; not our art and culture, but warning panels in languages destined to become extinct; not our technology, but its toxic residues.

Into Eternity is thus a haunting reminder of the troublesome material and cultural legacies of the first (perhaps, more accurately, the *early*) nuclear age. It suggests that a concern of the high Cold War, "containment" in all its forms, is not part of a defined (and finished) historical period, but is recapitulated as a long-term issue for nuclear culture.[3] If nuclear materials cannot be contained within the period in which they originate, if they leak into future eras, then neither nuclear issues nor nuclear culture can be safely consigned to, and contained within, an earlier historical period when global nuclear war seemed a possibility. Indeed, as nuclear industries, military and civilian, continue to prosper, nuclear materials and cultures do not merely persist, gradually attenuating over time, but proliferate. It might be that nuclear culture is less pressing on our general attention than it was during the Cold War, but it is not absent: slumbering and restlessly dormant, it is in fact troublesomely present. Just as Onkalo is predicated on the acknowledgement of long-term toxic threat, on the potential of leakage precipitated by either failure of containers (material deterioration; geological shift in the bedrock) or social change (ingression by those unaware of the danger, or desperate for the valuable resources—metals, etc.—it contains), so too is nuclear culture always threatening to reemerge into the contemporary moment.

Nuclear knowledge, and hence nuclear culture, is unidirectional: barring a civilization-ending global calamity, we cannot return to a state of knowledge entirely innocent of nuclear phenomena; nor, barring unforeseeable technological innovations, can we remove nuclear waste materials from the world or make them entirely safe. The challenge for Nuclear Criticism, whether we conceive of it as a specific theoretical approach, or simply as a critical interest in nuclear issues, is to mature beyond its Cold War adolescence and find a way to speak to long-term and more subtle manifestations of nuclear culture.

This chapter seeks to do two things. First, it traces the continued presence of nuclear markers in literature by discussing four texts published since the Cold War. Terry Tempest Williams' *Refuge* (1991) is both memoir and nature writing. It details threats to the Bear River Migratory Bird Refuge, where Williams worked, by the flooding of the Great Salt Lake in Utah, and it details the devastating impact of breast cancer, possibly produced by nuclear testing in Nevada in the 1950s, on the women in Williams' family. Bobbi Ann Mason's *An Atomic Romance* (2005), a novel in Mason's characteristic "K-mart realist" style, is also about the legacies of nuclear industry, building its story around a health scandal at a uranium enrichment plant. Lydia Millet's *Oh Pure and Radiant Heart* (2005) is a magical realist novel, imagining the consequences that follow when Manhattan Project scientists Robert Oppenheimer, Enrico Fermi and Leo Szilard reappear in twenty-first century New Mexico. Kamila Shamsie's *Burnt Shadows* (2009) is a saga following the intertwined fortunes of two families over three generations, beginning with the 1945 atomic attacks on Japan, ranging through post-war partition between India and Pakistan in the 1940s, and CIA involvement in funding Mujahedeen fighters in Afghanistan in the 1980s, and ending with events around the world resulting from the so-called "War on Terror" in the twenty-first century. The shadows of the novel's title are the images of cranes, heat-stenciled into the flesh of one of the protagonists, Hiroko, when the atomic flash at Nagasaki burns through her kimono.

Second, in discussing these texts, this chapter tentatively begins to work with some of the terms that might be important in the critical vocabulary of twenty-first century Nuclear Criticism. It draws on, and champions, critic Lawrence Buell's notion of "toxic discourse" and ethnologist Joseph Masco's conception of the "nuclear uncanny," but it also stresses the utility of the idea of "leakage," and understands the continuing manifestation of nuclear consciousness as brought about by a sort of "nuclear cultural osmosis."

The semipermeable membrane

The Cold War seems like another world now, its terrors and conflicts muted into sepia-tinged nostalgia by the passing of time. It ended so suddenly that its closing years, 1989-91, seem like an impermeable barrier between what was before and what came after and this dictates a particular narrative: before was the struggle between capitalism and communism; after was (in retrospect) the brief, happy hiatus of the 1990s and then, as the new century began, the move to an entirely other mode of geopolitical being with the World Trade Center attacks and all that followed. Before was a world under threat of global nuclear war; after was a world without that threat.

This narrative is, of course, fundamentally wrong. It implies, without quite stating, an absolute division between Cold War and post-Cold War. Yet its simplicity remains compelling: no-one is likely to accept it when they examine it directly, but it silently informs a lot of thinking. Frequently unspoken, the narrative underpins many assumptions, including the supposed superiority of neoliberal models of economy; the idea that, post-9/11, the terrorist threat imperils Western society more than anything that preceded it; and the assumption that, by and large, nuclear culture is a thing of the past.

I want to draw attention to just how leaky the seemingly impervious historical barrier at the end of the Cold War truly is. Certainly all of the literary texts to which this chapter refers point to a transgression of the boundary. Poised on the cusp between Cold War and post-Cold War, *Refuge* is a gateway text drawing attention to the flow of nuclear consequences into the future. Similarly, *An Atomic Romance* details the twenty-first century repercussions, for environment and health, of the Cold War military-industrial complex—a complex that persists in new forms in the contemporary moment. In *Burnt Shadows*, Hiroko carries her atomic signature, the cranes burnt into her flesh, from Nagasaki, through Tokyo, Delhi, Istanbul and Karachi, to New York City and the twenty-first century. In *Oh Pure and Radiant Heart* the 1945 selves of Oppenheimer, Fermi and Szilard wink into existence again on the other side of the Cold War, in twenty-first century New Mexico. Like nuclear materials, nuclear culture might trouble us with its leakiness.

This is not to say that nuclear culture is the same as it was during the Cold War. The force of nuclear culture is no longer present in issues that are an explicit and sustained focus for public discussion and it would be contrary to claim that the threat of the apocalyptic nuclear event has not receded. Rather, the impact of "the" nuclear emerges from an assumed body of background knowledge and beliefs (shared, of course, unevenly within and across populations) about how the universe is constituted, about the possibilities for human intervention in the world and about the nature of the environment itself. We can conceive of this having permeated the culture through a process of "nuclear osmosis." For the *Oxford English Dictionary*, osmosis is the "process by which molecules of water or another solvent tend to pass through a semipermeable membrane into a region of greater solute concentration, so as to make the concentrations on the two sides of the membrane more nearly equal" and (metaphorically) a "process resembling osmosis, *esp.* the gradual and often unconscious assimilation or transfer of ideas, knowledge, influences, etc."[4] Osmosis is hence a helpful metaphor with which to account for the dissemination and accumulation of cultural nuclear knowledge in a post-Cold War environment. Nuclear issues are not currently, on the whole, experienced as pressing and sustained; they are not at the fore of the cultural consciousness. Rather, they flicker into the cultural consciousness (as at Fukushima in March 2011; as in the drip-drip of news and feature stories about nuclear power, the history of the Cold War, future defense spending, etc.), accumulating in the cultural subconscious. Osmosis also, of course, might speak metaphorically (it is not the same as the process defined in the *OED*), to the process by which radiation and nuclear materials accumulate in and percolate through the environment. In both cases, there is a gradual accumulation of nuclear residues (material; cultural) that silently, as in the "silence" of fallout referred to in the title of this collection, build up without us being fully aware of them. These become apparent in moments of recognition when we catch glimpses of the full weight of the nuclear legacies (again: material *and* cultural) that continue to accrue throughout the world.

Contemporary Nuclear Criticism might, then, look for evidence of this accrual of nuclear material and knowledge in texts. Frequently, as in the examples discussed in this chapter, relevant texts involve moments of anagnorisis, when the full weight of the nuclearity of our world becomes apparent, but a Nuclear Criticism interested in the contemporary moment (in other words, one that is not oriented only to exploring Cold War nuclear cultural history) might equally be interested in passing, but largely unconscious, references that provide a clue to the, with apologies for the neologism, *nuclearity* of our world.[5] Critical concepts formulated by Buell and Masco are helpful in unpacking this nuclearity in action.

Toxic discourse

Wildlife refuges feature in important ways in both *Refuge* and *An Atomic Romance*. In the former, the Bear River Migratory Bird Refuge provides a "sanctuary" for Williams, a natural place apart from the anxieties of human life to which she can escape.[6] The latter begins with the protagonist, Reed, struck with an impulse to "run," and jumping on his motorbike to flee to the wilderness at Fort Wolf Wildlife Refuge.[7] In both instances, however, refuge rapidly ceases to be something that can be found in these supposedly wild places and in both instances nuclear materials are in some way implicated. A detour into the idea of toxic discourse might help to explain why.

Buell's essay, "Toxic Discourse" (1998), was an early intervention in the then emerging field of ecocriticism.[8] It distinguishes toxic discourse, both as an object of study and as a mode of reading, from (or at least as a distinct area within) that field, and the phrase points to the interpenetration of human and natural spaces. It thus rejects an ideal often held dear in nature writing of the pristine natural space—the absolute wilderness—although it acknowledges that such a space might remain a valuable ideal in environmental activism.[9] The implication of "Toxic Discourse" is the recognition of numerous vectors by which spaces interpenetrate each other.

A consequence we can draw from Buell's argument (though this is not quite the point he is making) is that conceptual boundaries we unthinkingly construct between nature and artifice are problematic. It is not that we should collapse them—to do so would be to render both concepts meaningless—but that we should more carefully question the distinctions we make between spaces valorized as natural and pure, and those seen as somehow artificial and contaminated. For example, we might draw from this reading of Buell's essay that we should move beyond a naively conceived rejection of the presence of "chemicals" in food in favor of "natural" products because such products are, of course, themselves chemically constituted. This does not mean we have to accept the presence of all contaminants in food, nor that we should assume manufacturing methods for producing processed foodstuffs are benign. What it does ask is for a more subtle understanding that acknowledges that what we assume to be "natural" and "artificial" are the poles of a conceptual spectrum, rather than entirely other to one another: between them lie complex areas where nature and artifice interpenetrate.

Although this suggests there is no pure nature to preserve, it should invigorate, rather than disable, environmental discourse. It asks that we hang our ecological aspirations on conceptions of a sustainable environment that acknowledge the co-presence of human and natural phenomena and agents. In this reading of the world there is no absolute natural space, no refuge from human involvement. We can still seek to filter our human/artificial incursions from various dimensions of, or spaces within, our "natural" world, but we must acknowledge that there is a broad integration between the two: the (natural) planet is a human planet.

Perhaps this is not quite the same point that Buell makes when he writes that toxic discourse insists on "the interdependence of ecocentric and anthropocentric values," but it shares toxic discourse's consequences which are to place the human and the social in the context of the natural world.[10] Buell illustrates his article with a series of examples, including Williams' *Refuge*, in which local communities become alive to the toxicity of their local natural and social environments. Toxicity provides a threat, but it also produces a sense of community through identification of shared peril and engagement in activism in response to that peril.

This detour into toxic discourse is relevant because it resonates with the ways in which Nuclear Criticism might reawaken us to the broad nuclearity of our world. It is not simply that, say, nuclear war or a nuclear accident could happen in the future (that our world is now non-nuclear, but may become so in the future), but that our world is already nuclear: it contains as naturally occurring phenomena materials (uranium, for example) and effects (radiation) we associate with nuclear industry. Furthermore, not only is our world materially different because of the twentieth century, but our perception of it is shaped by nuclear modes of understanding that we have inherited from the past. Our cultures are infused with this legacy; the nuclear is part of the warp and weft of our realities.

One of Buell's examples of toxic discourse is *Refuge*, but I want to expand on what he says in order to draw out the relevance of the text for contemporary Nuclear Criticism. *Refuge* is an important text: although it is rooted in Cold War experience, its double narrative points to long-term nuclear legacies. In part, it is a work of nature writing: chapters are named after birds, and the book closes with a list of birds associated with the Refuge; Williams discusses the pleasures to be gained from bird-watching, and the book is informative about birds' habits and characteristics. Each chapter also records, under its title, the current (to the narrative moment) height of the Great Salt Lake, thus charting how it rises and eventually floods. The book depicts local controversies about how to contain the flooding. As Buell points out, it is significant that the text moves between Great Salt Lake City, the Great Salt Lake and the Refuge because it points to the "inextricable imbrication of outback with metropolis."[11] Indeed, Williams explicitly draws attention to this co-dependence when the opening chapter begins with the driving directions for the twenty-five minute journey from her home to the lake. As she comments: "Great Salt Lake: wilderness adjacent to a city."[12] Natural and human worlds are intertwined.

Shadowing this narrative is another one, of personal loss. As Williams announces in the prologue, "Most of the women in my family are dead. Cancer. At thirty-four, I became the matriarch of my family."[13] The two narratives resonate with each other: the growing peril of the Refuge as the lake level rises mirrors the increasing threat as Williams' mother, Diane's, cancer advances, and questions about the desirability and efficacy of intervening to shape the course of natural events are raised by attempts to contain the flooding of the Great Salt Lake, and to pursue ever more aggressive medical interventions to contain the cancer.

Only in "The Clan of One-Breasted Women," the magnificent, angry essay that serves as epilogue to the text, does a nuclear context for all of this become apparent. As if providing power to a circuit, the epilogue electrifies many features already encountered in the main body of the narrative, bringing them to life in new ways. This hidden-then-revealed aspect of nuclear reality is a notable characteristic of post-Cold War representations of the material and cultural legacies of the early nuclear age.

Before it "becomes" a nuclear text, *Refuge* is something else, or rather some other things. Stylistically, it moves between memoir, nature writing and protest essay, its narrative instability perhaps an echo of the nuclear concerns that emerge at the end, for nuclear materials are unstable, mutating into different forms, and long-associated with, as Spencer Weart has pointed out, transmutation.[14] Indeed, we might conceive of a new reading of the nuclear as stressing precisely this constant "becoming": an identity coalescing and forming around a process of change.[15] The book's "becoming" nuclear in the closing essay is a shock: Williams notes factors that normally correlate with incidence of breast cancer (genetic predisposition; poor diet; childlessness; late pregnancy) and then throws in something new: "What they [the statistics] don't say is living in Utah may be the biggest risk factor of all."[16] Living in Utah is a risk because the state is downwind of the Nevada sites where atmospheric nuclear tests took place in the 1950s.

Williams discovers the connection when she recounts a recurring dream (a "flash of light in the night in the desert") to her father who reveals it has an origin in childhood experience: "You were sitting on

Diane's lap... It was an hour or so before dawn, when this explosion went off... [W]e saw it clearly, this golden-stemmed cloud, the mushroom... Within a few minutes, a light ash was raining on the car."[17] The encounter with the bomb, both with this particular one and with those that were more invisibly present, provides a possible explanation for the eruption of cancer in Williams' family: before 1960 only one had faced breast cancer; subsequently, nine women, seven of whom later died, had mastectomies (hence the clan of "one-breasted" women), and Williams herself had biopsies for breast cancer. (That it is not possible absolutely to pin down a causal link between the bomb and individual cancers is a crucial instability in the text to which I will return later.)

The dream codifies the contemporary nuclear experience as a return of the repressed and it is notable how *Refuge*'s nuclear concerns are with the delayed effects of something that has already happened rather than, as in most Cold War nuclear literature, with hypothetical future wars or incidents. However, as indicated above, it is not simply that "The Clan of One-Breasted Women" provides a possible explanation for the cancers that blight her family; it also invites us to read anew a whole host of previous incidents in the text, the significance of which was not previously apparent. It is not, therefore, so much that the text "becomes" nuclear, as that it becomes apparent that it was always already nuclear.

Nuclear references or incidents that appeared briefly and seemed peripheral—the description of a desert artwork as casting "a shadow across the salt flats like a mushroom cloud"; Williams' desire for the museum at which she works to sponsor a film about the disposal of uranium tailings; the cancelling of a hike on the day of an underground nuclear detonation at the Nevada test site; an outlandish plan to obviate the threat of flooding of the Great Salt Lake by "nuking" it to drain off water[18]—become retrospectively charged with a collective significance by the revelations in the epilogue. Faced, at the end, with the potentially devastating impact of local nuclear conditions, we are awoken to the common link between these disparate, and otherwise ephemeral, references. We become alive to the broader nuclearity of contemporary experience.

But it is not only what we might call this primary nuclear circuit that is electrified by the revelations at the end; secondary circuits are also lit up by it. For example, an incident recounted in the opening chapter of the book sees Williams visit the Refuge with a friend, Sandy Lopez. The women have a run in with men from the local gun club who have flattened the nests of burrowing owls. The nuclear context places the guns at the end of a spectrum of implements of warfare. In the 1980s anti-nuclear and pacifist movements had been invigorated by feminist activists, for whom the posturing of the Cold War nuclear standoff epitomized the consequences of patriarchy. Phallic nuclear missiles provided particularly suggestive symbols of a destructive culture of masculinity.[19]

Significantly, just before they encounter the gun club Williams and Lopez are talking "of rage. Of women and landscape. How our bodies and the body of the earth have been mined."[20] The exploitation of both land and women are continua in the broader workings of patriarchal oppression. The insistence on linking the human body with landscape is important for it suggests that we do not simply inhabit landscapes, but that we are part of them. Indeed, our bodies are themselves landscapes, as Williams notes when she writes that the "womb is the first landscape we inhabit."[21] So when both bodies and land are described as being mined this opens up space for contradictory metaphors of maternity and mining to be deployed to speak to the exploitation of both people and natural resources. "The Clan of One-Breasted Women" returns to this metaphor when Williams describes a dream of feminist protests about underground nuclear tests that are ongoing in Nevada. With each test, there is a "heave" in the desert and "[s]tretch marks appeared" as if the earth is trying to give birth. The women in the dream (later the essay describes an actual protest in which Williams participates) "couldn't bear it any longer. They were mothers. They had suffered labor pains but always under the promise of birth... [E]ach bomb became a stillborn."[22]

The text builds extensive connections between individual experience, including the intimate spaces of the body, and a broader nuclearized culture. One of the discoveries of the text is that refuge is not a place: the Migratory Bird Refuge, where Williams had gone for respite, floods, and she is forced into the discovery of a refuge rooted in psychological accommodation rather than physical escape. Indeed, seeming wilderness spaces are always problematized in the text. Williams discovers, for instance, that the desert she loves is a heavily militarized space. She comes across plywood tanks used as targets in firing ranges and is told that "[i]f you look straight up, that's not blue sky you see—that's military airspace."[23]

If we are to find refuge (and of course the term is extraordinarily loaded by a Cold War discourse where it referred to the domestic nuclear shelter) within a world that contains no true wilderness space (where human and natural worlds are intertwined, not separate), then it has to be in the form of an accommodation both to the compromising (the "toxicity," if you will) of natural space and to the existence of flux. Refuge becomes a psychological response to the world, not a space within it.

Like *Refuge*, *An Atomic Romance* opens up the always-already nuclearity of our world. Reed's journey on his bike to Fort Wolf Wildlife Refuge is, as I have discussed elsewhere, suffused with markers of atomic industry.[24] Clouds, for instance, that might otherwise signify pastoral retreat, seem to be produced by industrial chimneys.[25] As with *Refuge*, this is a text where an alertness to atomic contexts allows us to produce a nuclear reading that reveals both text and landscape in the act of becoming nuclear.

The place of refuge is itself also compromised, just as in Williams' text. When he is camping, Reed encounters "eerie blue flames," "lovely yet terrible" from a scrap heap near an industrial plant bordering the Refuge.[26] When it emerges that these may be linked to waste products from the uranium enrichment plant at which Reed works, he is forced to confront the violation of boundaries between this seemingly discrete natural space and local industry:

> There were no frogs or fish in the ponds; they had been killed to keep people from eating them… [H]ow could he have pretended this place was anything but a malignant jungle. He trembled with sadness—this violation, a crude intrusion into a natural place that he held to be sacred.[27]

Just as the "sanctuary" Williams finds in the Bear River Migratory Refuge is shattered, so too is the sanctity of the Fort Wolf Wildlife Refuge undermined for Reed.

Further, just as Williams and her family find that they are "nuclear," so too are Reed and his co-workers forced to confront their own bodies as nuclear spaces. Ed, Reed's uncle, tells him how he worked at the plant in the 1960s, "mixing black uranium dioxide and hydrofluoric acid to make greensalt," which got inside, and became part, of him: "I'd shower at work and I'd shower at home, but I couldn't get rid of it. I even spit green."[28] When Reed tracks down the results of health tests at the plant he finds his own nuclear legacy. Neptunium, declared in the figures, implies the presence of something more sinister and undeclared: "[He] realized that the body counter couldn't recognize plutonium, but if neptunium was in his body, then an equal amount of plutonium was there too, like a shadowy twin."[29] Like Williams, Reed makes the terrifying discovery that he is part of, not just existing in, a broader ecosystem. The broader psychological implications of the "shadowy twin" begin to take us into the issue of the "uncanny" double, to which I will return below.

Burnt Shadows and *Oh Pure and Radiant Heart* do not deal with health scandals or environmental contamination, but they participate in toxic discourse in more abstract ways. In both there is a leaching of nuclear issues, that we might expect to be contained by the end of the Cold War, through the historical boundaries that mark its end. In the former, Hiroko's presence in the twenty-first century United States, with the cranes resulting from the atomic flash burnt into her skin but concealed beneath her clothes, signals a return of nuclear consequences to their source. In the latter, the appearance in twenty-first century New Mexico of Oppenheimer, Fermi and Szilard, apparently the same age as they were at the moment of the Trinity test, signals the reactivation of ideas and movements that were lying latent since, rather than ended by, the close of the Cold War. Just as it is Williams' dream that marks the return of the repressed nuclear past in *Refuge*, so too are events begun in Millet's novel when Ann has a dream/vision about a man in the desert whom she later discovers to be Oppenheimer. The consequences of the atomic scientists' reappearance are to reactivate disturbing trends in nuclear culture. At Szilard's instigation, Oppenheimer leads a grand peace march on Washington, but the movement is hijacked by Christian fundamentalists and a strong nuclear apocalyptic strain reemerges as more and more followers take Oppenheimer's return to signal the coming Rapture.

We might thus understand toxic discourse, in its nuclear connotations, to be an elastic concept. It might, literally, signal the presence of, and a written interest in, toxicity, as in the nuclear agents that are the potential sources of health problems in *Refuge* and *An Atomic Romance*. However, serious though the problems highlighted by those agents are, most people are not made ill by contact with nuclear agents. We must thus, if we are to engage with the broader nuclearity of our world, engage with toxic discourse in a way that disconnects toxicity from its pejorative overtones. We might, for instance, maintain the position that our world "is" nuclear, without thinking it is so, for most people, in the sense of serious direct encounters with nuclear pathogens. Just as nuclear agents travel the vectors that interpenetrate ecosystems around the world, sometimes invisibly and sometimes with manifest effects, so nuclear ideas, and the idea of the nuclear itself, have travelled and suffused our cultural ecosystems. We can develop this idea by turning to the "nuclear uncanny."

The "nuclear uncanny"

Ann, through whom much of *Oh Pure and Radiant Heart* is focalized, lives not far, though she does not fully realize it initially, from one of the key birthplaces of the nuclear age. She and her husband, Ben, are "less than an hour's winding drive from the city of Los Alamos, on the high pink and gray mesa with its juniper trees and piñon bushes, salvia and chamisa."[30] The chamisa plant is the subject of a telling anecdote, about the return of the nuclear, in Joseph Masco's powerful ethnographic study of New Mexico, *The Nuclear Borderlands*. In Bayo Canyon, Masco writes, is buried a "nuclear waste treatment area, which was closed in 1963 ... and thought to be safely out of reach."[31] However, a chamisa plant, rooted on the site, sent "tap-roots fourteen feet into the earth... [where it mistook] strontium-90 for calcium, sucking it into its circulatory system and returning it to the earth's surface to reenter the food chain."[32] Masco claims it contains three-hundred thousand times more strontium-90 than a normal shrub.

For Masco, this is a "special creature of the nuclear age," and the fact that it thrives is an "example of the strange duality of the nuclear age, that contamination... can travel hand in hand with visible signs of health and prosperity."[33] This duality also makes the plant a symbol of the "uncanny," in the sense proposed by Freud in his 1919 essay ("Das Unheimliche") on the subject. For Freud, the "double" is particularly disturbing because of the way it straddles the boundary between the known and the unknown.

Drawing on Freud, Masco suggests that the nuclear age "has witnessed the apotheosis of the uncanny."[34] This is partly the result of Cold War nuclear and missile technologies which, by threatening the delivery of nuclear war anywhere in the world with little or no warning, effectively collapsed our sense of the pace at which the world could transmute into an entirely other and alien place. More significantly though, for this chapter nuclear materials are "uncanny" in Masco's analysis because of their material qualities. As radiation is "colorless and odorless, yet capable of affecting living beings at the genetic level... nuclear materials produce the "uncanny" effect of blurring the distinction between the animate and the inanimate, and between the natural and the supernatural."[35] This is the "uncanny" in Freud's sense of dislocating selfhood by rendering the experience of reality as "other" to itself; as similar, but slightly different; as profoundly alien precisely because it seems so nearly familiar.

While Buell's "toxic discourse" is suggestive of the ways in which the spaces of the world are compromised by toxicity (and, stretching Buell's term, how our psychic and cultural spaces are colonized by the nuclear), Masco's "nuclear uncanny" provides a productive way to engage with the profound impact this can have on our sense of self. As Masco says, "[o]ne psychosocial effect of nuclear materials is to render everyday life strange, to shift how individuals experience a tactile relationship to their immediate environment."[36]

Consciousness of toxicity means not only that spaces are, as in Buell's analysis, seen as "other," but so too is the self. As Buell himself points out, the "modern nature that toxic discourse recognizes as the physical environment humans actually inhabit is... a network or networks within which... humans are biotically imbricated."[37] This demands an understanding of the self not as self-contained, nor even as part of the environment, but *as* an environment itself, or, more accurately, one among many subsystems of a larger environment and itself consisting of smaller sub-environments. Through materials we ingest and excrete, through the air we breathe, indeed through those agents for which the skin is not a barrier but a permeable membrane, we are giving to, taking from, and constituted by the environments around us. Individual selfhood does not disappear as such, but it shimmers with uncertainty around the edges; it is rendered problematic; it is a form (in space; in our mind), not a thing. Nuclear agents are thus not outside, necessarily, but inside, constitutive of us as much as they are alien to us. We might think of nuclear culture—of nuclearity, of the experience of being nuclear—in a similar way.

When the nuclear scientists turn up in *Oh Pure and Radiant Heart* they do not so much bring anything new to the twenty-first century, as reveal how nuclear it already is: it looks the same, but we understand it differently. From Oppenheimer's point of view, having jumped six decades, New Mexico is an "uncanny" world: "'It's worse than a different world,' he says, 'it's the same world turned alien'."[38] Ann herself is changed in a similarly "uncanny" way by her encounter with the nuclear scientists. Skeptical of their claims to be who they are, Ben, her husband, finds himself seeing Ann, who believes in them, differently. When he kisses her, she has become strange through her involvement with them:

> She was so known, the sweet smell of cheeks and the nubs of elbows, rough skin over the smooth and nosy bone... He knew exactly the span of her wrists, the angle of her chest and shoulders learning in to him, how her weight felt different from others and no other weight could ever feel like hers by mistake. The same as ever, except for her insistence on the impossible, everything as ever except what screamed *never before*.[39]

She is the same as before, yet "other" to that person whom he loves.

The presence of the "uncanny" in *Oh Pure and Radiant Heart* signals the return of the repressed: the repressed nuclearity of our world. Structurally, the novel signals this by interspersing through its narrative of the post-Cold War present anecdotes and information from the Cold War (and, by the end of the book, post-Cold War) past. These moments of nuclear history and culture bubble into the text, usually separated off only by white-line breaks, but sometimes not even by these. Sometimes a new paragraph will simply switch, without warning, from the present-day narrative to, say, a quotation from an eight-year-old girl who escaped Hiroshima.[40] These things are part of our present, not buried safely in the past, the novel implies.

For *Oh Pure and Radiant Heart*, the nuclearity of our world lies in three things. First, there is an entrenched (if latent) nuclear culture, brought into visibility by the return of the scientists. Second, the world is materially constituted by nuclear materials and in particular the residue of nuclear industry. For example, one of the nuclear "anecdotes" mentioned above works to educate us about the presence of this nuclear material in our world:

> Plutonium-239... has often been called "the deadliest substance known to man." Yet in the years since Trinity, underground nuclear tests have left more than eight thousand pounds of plutonium in the ground, while aboveground tests—with a total yield roughly equivalent to twenty-nine thousand Hiroshimas—have put at least nine thousand pounds into the atmosphere.[41]

Third, the novel suggests, the world is politically and militarily geared up for nuclear industry, and much of its space is nuclear, if we could only learn to see it. The novel claims that "nuclear weapons production facilities occupied over three thousand square miles of US territory" at the turn of the century.[42] At the end of the novel, when the scientists have disappeared again, we are told that "[t]he bombs they had conceived remained, of course; the bombs in their various silos, trucks and trains, their submarines and aircraft, had been dispersed over the globe like seeds, and lay quietly waiting to bloom."[43] The mixing of pastoral imagery into this lurking technological future is another example of the "nuclear uncanny" and echoes the toxic discourses we have already encountered in *Refuge* and *An Atomic Romance*.

The novel therefore suggests a continuity between the military-industrial complexes of the Cold War and post-Cold War periods. A similar sensibility is present in *Burnt Shadows*. The "War on Terror," the context for the final section of the book, is presented as a recapitulation of earlier anxieties, not as an entirely new age. So, when Hiroko walks the streets of New York after the World Trade Center attacks, the notes pinned up by people searching for lost loved ones bring to mind the "train station at Nagasaki... The walls plastered with signs asking for news of missing people."[44] Hiroko has returned to the very source of the nuclear attacks on Japan (we might note that she is in New York City and that it was the

Manhattan Project that produced the bomb that destroyed her home). She serves to unsettle our sense that the atomic past is finished with. "A week after India's nuclear tests, with Pakistan's response in kind looming," we are told, "she didn't see the ache in her back as a result of the long plane ride but rather a sign of her birds' [the cranes', burnt into her flesh] displeasure that she should have chosen this of all countries, as her place of refuge from a nuclear world."[45] Once again, we encounter the absence of refuge from the nuclear.

Hiroko makes legible the nuclearity of our world by her presence. She estranges us from our sense that the world is no longer nuclear. Indeed, she is a symbol of the impossibility of escaping the past. For instance, although she speaks with anger about how the term "Hibakusha" (explosion-affected person), with which survivors like herself were labeled, "reduces you to the bomb. Every atom of you,"[46] not only she but also her son are destined to carry this mark with them. Her son, Raza, is told that "[e]veryone knows about your mother... No one will give their daughter to you in marriage unless they're desperate, Raza. You could be deformed. How do we know you're not?"[47] The mark here is psychological or cultural more than it is physiological: Raza is perceived to be toxic, and this is almost more important than if he actually is. Again, we are presented with the disturbing image of the "double," of the "uncanny." Raza looks no different, whether he "is" nuclear or not, but his family history changes everything about how he is seen.

Similar dislocations of the self are present in *An Atomic Romance*. Reed becomes increasingly conscious that he may "be" nuclear because of his contact with nuclear materials through his work. Although he laughs off his lover Julia's concerns at the beginning of the novel ("I've been living with that stuff so long my insides would be neon green if you opened me up"),[48] his attitude shifts as more and more revelations about the plant emerge. We never discover if he will be made sick, but his attitude to his body changes. When Julia says that "[p]lutonium collects in the bones," Reed appears unconcerned, but later begins to distrust his body and not quite see it as his own.[49]

Refuge, of course, does deal with sickness as a (possible) result of nuclear contamination. It marks a changed attitude to the self through disturbing reworkings of imagery of female sexuality and maternity. The absence of a breast in "The Clan of One-Breasted Women" is both an illustration of the damage done to women by contemporary techno-culture and a marker of a new identity. The experience of cancer is conveyed as a making strange of the body to itself. Mimi, Williams' grandmother, for instance, is described as having a "distended belly, pregnant with tumor," and there is a disturbing scene in which she feels "labor pains" in the night, goes to the bathroom and gives "birth to a tumor."[50] This scene marks not just the horrific psychological accommodations people have to make as their bodies change under the influence of cancer—indeed, *become* cancerous—but a wider unsettling of the distinctions between what is natural and unnatural in the text. On the one hand, cancer is unnatural, the birth of new cells proliferating uncontrollably and changing the body; on the other hand, it is one of the natural ways by which bodies change and die.

Although the role of nuclear testing might seem to mark the cancers as unnatural, Williams is aware that there is no way of establishing a direct causal link. *Refuge* is beset by this uncertainty. As Buell comments, it has to employ a "discourse of allegation rather than of proof" and Williams comments that she "cannot prove that my mother, Diane Tempest Williams, or my grandmothers, Lettie Romney Dixon and Kathryn Blackett Tempest, along with my aunts developed cancer from nuclear fallout in Utah."[51] There is no way to trace an individual incidence of cancer back to nuclear testing, even though rises in the overall rates of cancer are a probable (if one shares Williams' reading of the evidence) result of testing.

The "toxic" presence of nuclear materials must then be understood as a psychological as well as a material phenomenon. It is about seeing the world differently; seeing how it is made strange to us—rendered "uncanny"—when we become conscious that our culture is, amongst other things, an inherently nuclear one. Our culture is produced as nuclear not only by the presence of contemporary nuclear industry, but through leakages—of materials certainly, but also of narratives and concepts—from our shared nuclear past. Leakage is bound up with the nuclear: in so many of its forms, nuclear materials and culture flout spatial and temporal containment. Our world's nuclearity is thus not fixed; it is an unstable phenomenon, shifting and reinventing itself, constantly becoming (and in the threat of nuclear disaster, threatening to become) something else.

Conclusion

Madsen's film about Onkalo, with which this paper started, demonstrates how our legacy wastes are intrinsic to our future. While Madsen raises intriguing questions about how we deal with nuclear waste, his film is sympathetic to the planners and engineers who are constructing Onkalo. They wrestle, like him, with the difficult social and cultural questions nuclear waste poses. The existence of the facility at Onkalo does, at least, show a willingness to engage directly with the issue of long-term storage, rather than to postpone this engagement and hope for the best.

We might, nevertheless, read the facility as a symbol of our desire to bury, to contain, not only nuclear material but also nuclear culture. As the texts with which this chapter deals demonstrate, such repressions have a tendency to return. Nuclear Criticism therefore has an important role to play, not only in continuing to open up nuclear culture of the Cold War period (reading the Cold War as a cultural category as well as a geopolitical reality of the late twentieth century), but also to expose how the twenty-first century world remains nuclear. Our world is different than it would otherwise be because of its nuclear past; it has a complex nuclear present; and it is subject to numerous potential nuclear futures. Nuclear Criticism's role is to map these changing nuclear worlds.

Notes

[1] "Legacy waste," a preoccupation of Bobbie Ann Mason's *An Atomic Romance*, discussed later in the chapter, refers both to the legacy of waste we inherit from previous generations and that we bequeath to those who come after us.
[2] *Into Eternity*, directed by Michael Madsen (2010; London: Dogwoof, 2011), DVD.
[3] We should note that "containment culture" originates in the specific geopolitical context of US "containment" of communism in the early Cold War and so means something different if we use the phrase to refer to later periods. For pioneering analyses of containment culture in the early Cold War see Alan Nadel, *Containment Culture: American Narratives, Postmodernism, and the Atomic Age* (Durham: Duke University Press, 1995) and Elaine Tyler May, *Homeward Bound: American Families in the Cold War Era*, 2nd ed. (New York: Basic Books, 1999).
[4] *Oxford English Dictionary*, s.v. "osmosis," accessed July 2, 2012, www.oed.com.
[5] The *OED* defines "nuclearity" in two ways: in a chemical sense, and in the sense of being at the center of things; in other words, not in terms of nuclear materials. *Oxford English Dictionary*, s.v. "nuclearity," accessed July 2, 2012, www.oed.com. I deploy the term to mean a prevalence and intensity of nuclear culture and experience.
[6] Terry Tempest Williams, *Refuge*, 2nd ed. (New York: Vintage, 2001), 19.
[7] Bobbie Ann Mason, *An Atomic Romance* (New York: Random House, 2006), 3.
[8] In literary studies, ecocriticism is a critical interest in the representation of the environment in literature, and in the intersections between the natural and the textual. The term encompasses a range of theoretical orientations, loosely linked in their commitment to a "green" agenda.
[9] Laurence Buell, "Toxic Discourse," *Critical Inquiry* (1998): 656.
[10] Ibid., 639.

[11] Ibid., 659.
[12] Williams, *Refuge*, 5.
[13] Ibid., 3.
[14] Spencer R. Weart, *Nuclear Fear: A History of Images* (Cambridge: Harvard University Press, 1988), 401-06.
[15] I am grateful to Morgan Shipley and Michael Blouin for this suggestion of nuclear becoming as a new reading of the nuclear.
[16] Williams, *Refuge*, 281.
[17] Ibid., 282-3.
[18] Ibid., 127, 44, 134, 248.
[19] See, for instance, Helen Caldicott, *Missile Envy: The Arms Race and Nuclear War*, rev. ed. (Toronto: Bantam, 1986).
[20] Williams, *Refuge*, 10.
[21] Ibid., 50.
[22] Ibid., 288.
[23] Ibid., 185-6.
[24] Daniel Cordle, *States of Suspense: The Nuclear Age, Postmodernism and United States Fiction and Prose* (Manchester: Manchester University Press, 2008), 148.
[25] Bobbie Ann Mason, *An Atomic Romance* (New York: Random House, 2006), 4.
[26] Ibid., 10.
[27] Williams, *Refuge*, 207.
[28] Mason, *Atomic Romance*, 31.
[29] Ibid., 178,
[30] Lydia Millet, *Oh Pure and Radiant Heart* (London: Vintage, 2007), 13.
[31] Joseph Masco, *The Nuclear Borderlands: The Manhattan Project in Post-Cold War New Mexico* (Princeton: Princeton University Press, 2006), 33.
[32] Ibid., 33.
[33] Ibid., 32, 33.
[34] Ibid., 27.
[35] Ibid., 30.
[36] Ibid., 33.
[37] Buell, "Toxic Discourse," 657.
[38] Millet, *Oh Pure and Radiant Heart*, 121.
[39] Ibid., 95.
[40] Ibid., 215.
[41] Ibid., 274.
[42] Ibid., 274.
[43] Ibid., 453.
[44] Kamila Shamsie, *Burnt Shadows* (London: Bloomsbury, 2009), 274.
[45] Ibid., 287.
[46] Ibid., 100. The use of the word "atom" is not, of course, innocent of its broader implications.
[47] Ibid., 189.
[48] Mason, *Atomic Romance*, 12.
[49] Ibid., 148.

[50] Williams, *Refuge*, 263, 244.
[51] Buell, "Toxic Discourse," 659; Williams, *Refuge*, 286.

CHAPTER TWELVE

IN A DARK WUD: METAPHORS, NARRATIVES, AND NUCLEAR WEAPONS

JOHN CANADAY

In 1998, Christopher Felver released a short film celebrating the sculptor Donald Judd. Footage of Judd talking about his work alternates with commentary by a poet, John Yau, who begins the film with these words:

> [Judd] would be against metaphor because he would understand, instantly I think, that metaphor could not in any way be a vehicle of meaning after the Holocaust and the dropping of the atom bombs; that metaphor had been called into serious question and serious doubt; and that he had to find a way to make art that could exist on its own terms without the support of metaphor, one may say, without the scaffolding of metaphor.[1]

Yau may be thinking of Theodor Adorno's famous dictum, "To write poetry after Auschwitz is barbaric," though he substitutes "metaphor" for "poetry," which transforms Adorno's highly debatable claim (he more or less retracted it in 1966[2]) into pseudo-intellectual folderol. Evidence of the continuing vitality of metaphor since 1945—in art, in literature, in popular culture, even in Yau's own work—belies the claim. Yet Yau's assertion, like Adorno's, points to something real. The cataclysms of World War II haunt the creative imagination. Urgency and inadequacy twine in us, along with a sense that something essential in the human condition has changed. Though metaphor remains a powerful means of expressing many different kinds of meanings, can it in fact help us to understand the bombings of Hiroshima and Nagasaki? In what ways might it help us to envision, and ultimately to create, a livable future in the long shadow of those events?[3]

Why might someone believe that the bombings of Hiroshima and Nagasaki had negated the ability of metaphors to convey meaning? It is the sweeping character of Yau's assertion that robs it of sense. If we put it in less drastic terms—if we suggest, for instance, that metaphorical representations can seem inappropriate, inadequate, even distasteful in the face of the suffering caused by the use of nuclear weapons—Yau would not be alone in experiencing a degree of anxiety. Writers who use metaphors to describe the bomb often seem uncomfortable with them. They try to limit their scope and control their reach, striving, apparently, to produce tracts of clear meaning—but at what expense? And what causes this anxiety? I will begin by considering the problem of representing the bombings of Hiroshima and Nagasaki before turning to attributes of the weapons themselves.

Challenges of Representing Nuclear Weapons

The most immediate challenge involved in representing the bombings of Hiroshima and Nagasaki is obviously the scale of human suffering they caused. How can we convey the immensity of the horror? The fullness of a single victim's experience is beyond our representational reach; multiply that by hundreds of thousands; string it out over years, over decades, in all its agonizing shadings: the task is beyond us. Yet this does not mean we should (or can) abandon it. Works like Masuji Ibuse's *Black Rain* (1965) and John Hersey's *Hiroshima* (1946) demonstrate the power and importance of such attempts. Even if any single work must fail to capture the scale of these events, each adds to our collective effort to open ourselves, to reach out, to understand.

Most people who attempt to write about these events are also hampered by a lack of direct experience. Without first-hand knowledge, we must rely on the accounts of those who have such experience. From this position of dependence, what can we add to eyewitness accounts? And why should readers trust the authority of someone who has not directly observed what he or she writes about? These questions are, in a sense, the motivation for this chapter, and it can be read as an extended attempt to answer them. For the moment, however, I will simply suggest that although eyewitness accounts will always be a crucial resource—and we should return to them often, as touchstones for our work—our task is to grapple with these events on behalf of readers who, like us, have no direct experience of nuclear weapons. For it is important to understand not only what these weapons meant to those who built and used and suffered them, but what they mean to all of us, in our own disparate contexts.

Why would we want to undertake this difficult work? Concern for the future of humanity seems a compelling answer. Yet there is hesitation even here, in the midst of this "noble undertaking," because in our efforts to grapple with history and make it speak to our future, we are acting in an essentially selfish fashion. There is immense presumption involved in appropriating the experiences, the suffering, the lives and deaths of the inhabitants of Hiroshima and Nagasaki. By what right do we turn them to our own purposes? Though most people (the *hibakusha* foremost) will insist that we must redeem the horror by learning from it, anyone who has attempted to do this will recognize the danger. We walk a fine line between honoring the dead and misunderstanding, misrepresenting, misappropriating their experiences. We must approach the task cautiously, respectfully. Yet too much caution can hobble our efforts, cause us to second-guess ourselves, prevent us from the necessary daring. The endeavor requires an artful balance of humility and hubris.

In addition to the challenges of representing the bombings of Hiroshima and Nagasaki, there are a number of peculiar features associated with nuclear weapons themselves that would seem to set them apart from other kinds of weapons, potentially rendering them more difficult to represent. As the Fat Man and Little Boy bombs demonstrated, the magnitude of their destructive force dwarfs conventional weapons. With the development of fusion bombs, the yield of nuclear weapons increased even further. The United States has produced bombs with a yield-to-weight ration approaching 6 kilotons TNT equivalent per kilogram.[4] This is six million times the energy release of TNT. Likewise, the largest nuclear weapon ever tested, the Soviet Union's Tsar Bomba, produced a yield equal to 50 million tons of TNT—comparable to two and a half thousand Fat Man bombs. Such destructive power defies self-defense and overwhelms imagination, frustrating our descriptive efforts; perhaps this is why some people assume it also cripples our modes of description themselves.

Nuclear weapons not only undermine the hope that we can defend ourselves, they also unsettle confidence in our ability even to detect an approaching threat. Hiroshima and Nagasaki were each destroyed by a single bomb carried by a solitary aircraft.[5] Today, a single Trident submarine carries 24 Trident II (D5) ballistic missiles, each with a range of more than 4,600 miles, and each capable of carrying eight independently targeted warheads, such as the 475 kiloton W88 (with an explosive force greater than 23 Fat Man bombs).[6] This means each Trident submarine can carry the equivalent of over 91 million tons of TNT (four and a half thousand Fat Man bombs) and is capable of utterly destroying 192 cities. Since Trident missiles travel at approximately 4 miles per second, they can reach targets 4,000 miles away in about 20 minutes.[7] The apparent speed and ease with which nuclear weapons can reach and destroy any location on earth is indeed terrifying.[8]

With the development of nuclear weapons came the realization that we now have power of an entirely new kind: the means to destroy ourselves, to destroy, indeed, the entire world. At the height of the Cold War, the United States and the Soviet Union had approximately 70,000 warheads between them; since then, the numbers have lessened dramatically. Today there are approximately 26,000 – 30,000 nuclear warheads in the world.[9] Before we congratulate ourselves on our improved safety, however, we should recall that this represents the equivalent of approximately 5 billion tons of TNT, or roughly 250,000 Fat Man bombs—more than enough to do ourselves in.[10] If describing the bombings of Hiroshima and Nagasaki tests our representational resources, how are we to cope with the prospect of a quarter million such cataclysms? And beyond that, how are we to imagine and embody in language not only our own end but the end of our species?

A more subtle but no less daunting challenge to our representational methods is the fact that nuclear weapons employ a recently discovered and "mysterious" source of energy. Most people do not understand how these weapons work—nor even, despite exposure to these concepts in high school science classes, how atoms are structured. Even with such knowledge, describing subatomic processes is immensely difficult, simply because they are beyond our ability to observe directly. Indeed, not only are they too small for us to see, but they do not, in fact, "look" like anything.[11] Although we are made up of them, atoms are entirely outside our experience of the world, behaving in ways that defy common sense. Scientists struggle with this difficulty as they seek ways of describing entities that are inaccessible to our senses and have no consistent analogs in the human world.[12] This would seem to be a considerable obstacle confronting anyone who wants to write about nuclear weapons.

In addition to the explosive release of energy, the physical processes underlying nuclear weapons have another threatening and even more mysterious manifestation: radioactivity. Our senses cannot detect it, yet it can kill us. Its effects can be immediate or delayed. It can cause hair loss, cataracts, internal bleeding, bruising, fatigue, weakness, nausea, blistering. It can linger in the body, cropping up years after initial exposure, resulting in cancer, infertility, organ failure. Radioactive debris can permeate the ground, rendering it dangerous long after the initial detonation of a bomb. Plants, in turn, can absorb the contamination, rendering the grains we eat toxic, poisoning our livestock and tainting their milk. Radioactive dust can rise into the atmosphere, descending as a deadly rain of fallout on distant places. Not only are these phenomena horrifying, but they present their own challenges to our representational methods.

Ironically, nuclear weapons are a particular manifestation of physical processes that also have beneficial applications. In fact, benign uses of nuclear technology appeared long before the malign,[13] and development of humanitarian applications continues side by side with nuclear weapons research. Nuclear power plants, for instance, offer the potential of vast, long-term sources of energy that—despite important concerns regarding operating safety, expense, and the creation of hazardous waste—merit serious consideration, especially in light of the environmental threats posed by global warming. Radiation treatments, likewise, have revolutionized medical practices, while radiochemistry has extended our understanding of biological processes, opening up new possibilities in the study of DNA and other molecular structures. There are potentially even productive applications of bomb technology, such as the possibility of breaking apart large meteors before they hit the earth. Yet despite the benefits of these applications of nuclear technology, they, like nuclear weapons, though more subtly, are also attended by a good deal of anxiety—as anyone who has undergone radiation treatments, or lived near a nuclear power plant, or had a loved one who worked in radiation research can testify.

Responses to Representational Challenges

Nuclear weapons clearly involve significant challenges to our representational efforts. I think it is important to recognize, however, that none of these challenges—on its own—is quite as unprecedented as we often allow ourselves to assume. The enormous destruction caused by the bombings of Hiroshima and Nagasaki was prefigured, for instance, in the fire bombings of Dresden, Hamburg, and Tokyo. The potential for surprise involved in the use of nuclear weapons echoes the attack on Pearl Harbor. The ability to destroy the planet—or at least render it uninhabitable by humans—exists in the form of global warming, to which each of us contributes every day. The mysterious character of nuclear processes is hardly more of an impediment to our understanding than larger scale atomic interactions, since most people do not understand how conventional explosives (including handguns) work any more than nuclear ones. The insidious threat of radioactivity is no more difficult to detect and defend ourselves against than the innumerable carcinogens and pathogens, both natural and artificial, that surround us (and pass through us) every day. Though each of these analogous examples test our representational abilities, none make us fret that metaphors simply don't work the way they used to.

Of course there are differences inherent in each of these comparisons. The bombings of Hiroshima and Nagasaki *were* different from the attacks on Dresden, Hamburg, Tokyo, and Pearl Harbor. Nuclear annihilation *is* distinct from global warming, nuclear reactions from chemical reactions, radiation poisoning from cancer. But none of these differences are sufficient to account for a belief that nuclear weapons cripple metaphorical representation while the analogous instances I have mentioned don't. Only in combination do the phenomena associated with nuclear weapons constitute something "new"—only together do they provide a satisfactory explanation for the almost divine powers we have ascribed to nuclear weapons.

The difficulty involved in representing the range and variety (and, of course, the power) of these attributes helps explain our complex, often contradictory responses to nuclear weapons. They are variously interpreted as promising the end of war and as threatening the end of the world; as diabolical devices and as agents of divine will; as practical tools of war and as immoral weapons of genocide; as mere technological artifacts and as proof of the epistemological supremacy of science. Contradictory attitudes towards nuclear energy permeate post-war culture, appearing in the public pronouncements of scientists, in political debates, in the manufacture of commodities, and in the full range of popular entertainments, from music to movies to pulp fiction. Everywhere we look, we can see ourselves struggling to make sense of the forces—physical and social—embodied in nuclear weapons: the dark humor of *Dr. Strangelove* and the macho slapstick of *True Lies*; Joseph Rotblat's Pugwash conferences and Edward Teller's response to Three Mile Island; Atomic Fireball candies and Brumm's Fat Man and Little Boy toys; Radioactive Records and the song "5 Minutes" by Bonzo Goes To Washington; Raymond Brigg's *Where the Wind Blows* and Russell Hoban's *Riddley Walker*; the Bulletin of the Atomic Scientists and Los Alamos National Laboratory. Instances of such ambivalence toward nuclear weapons are widespread, and they have proven remarkably resistant to overt attempts at logical resolution as well as to the implicit forces of cultural consensus.

Given the range and incongruity of the characteristics we associate with nuclear weapons, it is not surprising that writers often feel the inadequacy of any particular metaphor, preferring the relative safety of literal narratives. Though narratives also feel inadequate—unable to convey the immensity of the subject, or even fully account for a given individual's experience of it—they can claim a factual grounding in events (if only those of a fallible individual). Metaphors, on the other hand, though used in both fiction and nonfiction contexts, have a fictional, aesthetic feel to them. They impress us as the willful creations of a writer. In *Hiroshima Diary*, for instance, Michihiko Hachiya describes a "bundle" of medical supplies as "no bigger than the tears of a sparrow."[14] This simile stands out as a rare instance of authorial license: in the midst of the terrors and suffering Hachiya describes, this shift into self-conscious linguistic play feels odd, almost disturbing. A part of us whispers: *these things are too awful to make art out of.* Yet the straightforward narrative mode that characterizes the book, founded on Hachiya's direct experience of the events he describes, gives him the authority to indulge in this moment of authorial creativity.

But is metaphor the decorative trifle that this attitude suggests? Or can it be a medium for the serious exploration of human experience? Given that our society valorizes the processes of logical analysis and causal narrative—in science, in the social "sciences," in academic discourse, and even in our interactions with one another—can metaphorical representation complement rather than negate these dominant modes? If so, then it is imperative that we "make art" out of precisely those experiences we deem most important and least accessible to human reason. Answering these questions requires that we consider carefully the dynamics of metaphorical representation—how it works and what it offers us.

How Metaphor Works

Metaphors are linkages between symbols—though we generally ignore the intermediary role of the symbols and focus instead on the links between what the symbols represent. Thus in Hachiya's comparison of medical supplies and sparrow tears, we do not think of the metaphor as a linking of symbols (the actual words "bundle" and "tears") but of the objects to which those symbols point. Though this is a simplification of the actual processes involved in metaphorical uses of language, it is a useful one for our purposes—and it is so common as to be unavoidable in any case.

The ostensible purpose of a metaphor is to convey understanding of an unfamiliar object, event, or idea (the tenor) by linking it to something familiar (the vehicle). For instance, "Mary is a crab," tells us something new about Mary by equating her with a familiar animal. These linkages can take the form of comparisons, equations, or replacements: "Mary is acting like a crab"; "Mary is a crab"; "The crab came by today." Metaphors used as overt comparisons tend to be the most highly circumscribed, in several senses: they make the act of linking obvious by means of comparative terms ("like," "as"); they link less forcibly, stopping short of asserting actual identity; and they often narrow the focus of the linkage (in this case by identifying behavior as the basis of similarity). Replacement metaphors, on the other hand, are the most intensive, far-reaching, and "poetic."

As this suggests, an essential feature of these linkages is that they involve multiple comparisons and contrasts in a single verbal gesture. Though the writer usually has a small number of similarities in mind when formulating a metaphor, the actual link involves a practically unlimited number of connections. Mary might look like a crab in any number of ways, from beady eyes to thin, bony arms. She might move in a crablike fashion. She might behave aggressively. She might like pinching people. She might maintain a strong social "shell," rendering her difficult to get to know. She may be a scavenger. She might prefer to live by the shore. Though some of these possibilities will seem more likely than others, the relative likelihood is a function of social conventions and prior uses. Since each of us has encountered and absorbed these precedents differently, any given metaphor is "open to interpretation" along a wide variety of lines, depending on which threads resonate with a given individual's particular memetic makeup.[15]

Metaphors therefore induce a curious state in a reader, requiring the recognition, evaluation, selection, and rejection of connecting threads. This is work each of us does all day, every day, without "thinking" about it—at least, not consciously. Often the process is fairly straightforward: social conventions and repetition vet popular metaphors ahead of time. When someone says, "I left the car running," she does not pause to select the metaphor, nor does anyone listening expend energy considering in which ways a car might be "running" or what the comparison is intended to convey about the state of the car. Metaphors of this sort have become static, their multiple semiotic threads snipped and woven into a single strand of meaning. Many metaphors, however, remain dynamic even after repeated use—and all "original" metaphors require active interpretation. In what ways is Mary like a crab? Processing this metaphor—retrieving and unpacking the meaning it is meant to convey—requires an active mental engagement involving a complex set of steps. Luckily, our brains have been wired to perform these tasks quickly and efficiently, as is evident in the fact that we spend relatively little time thinking consciously about the myriad metaphors we encounter every day.

Not surprisingly, despite our rather astonishing metaphorical skills, we generally sacrifice some precision in favor of speed. Told that Mary is a crab, we don't actually consider all the possible bases of comparison. Instead, we quickly eliminate large categories, narrow in on others, and apply the gist of the similarity, as we interpreted it, to our ever-flowing "stream of thought."[16] As we might expect in such a seat-of-the-pants process, mistakes are often made and may later be discovered: "Oh, you meant the way she walks! I thought you meant she was irritable." Yet mistaking a speaker's intent is only one way in which metaphors can surprise us. Even after a metaphor has been processed and its meaning "extracted," innumerable rejected or overlooked possibilities remain, and seemingly static threads can regain their dynamism in surprising or significant ways. Even the framer of a metaphor can discover unforeseen and unexpectedly apropos threads, and, in some important senses, the precise contours of a metaphor's meanings are never firmly fixed.

Furthermore, metaphors are not unidirectional linkages. Although the ostensible purpose of a metaphor is to increase our understanding of the tenor, the comparison ultimately goes both ways. It is a given that the vehicle must be familiar in order to convey information; yet the tenor cannot be entirely unknown, either. To say, "X is a crab," is not a metaphor. It gives us no basis for discrimination. We must map *all* the qualities of a crab onto the unknown noun. For all we know, it *is* a crab. This suggests, of course, that the differences between the vehicle and tenor of a metaphor are as important as the similarities. It also points out that metaphors are only viable when it is possible to map the properties of the tenor back onto the vehicle, if only as a means of determining which of the vehicle's properties may be appropriately mapped onto the tenor. Practically, it will often be the case that we will have a much more detailed knowledge of the tenor than of the vehicle: the tenor is, after all, the primary object of interest. In the case of our working example, for instance, we are likely to know a lot more about Mary in particular than *Brachyura* in general.

Literary uses of metaphors explore these properties and turn them to the service of expressing the meanings of human experience in non-linear ways. Literary metaphors, for instance, tend to resist our efforts to narrow their scope and, instead, keep a wide variety of meanings in play at once. They also emphasize semiotic mapping in both directions, from tenor to vehicle as well as from vehicle to tenor. In short, they create webs of meaning—as opposed to the linear sequences generated by the causal structures of narratives. When, for example, Eleanor Wilner writes in "High Noon at Los Alamos" that "our thoughts, however elegant, were fire," she uses the metaphor to map a number of properties associated with fire onto human thought: energy, light, beauty, speed, the consumption of whatever fuels it, its destructive capacity.[17] But the metaphor also maps the other way, creating what is, in this case, perhaps an even more powerful and disturbing link. Our thoughts not only share some of the qualities of fire, Wilner implies, they also stoke it, spread it, manufacture it. Our thoughts *become* fire—from the burning of Troy to the bombings of Hiroshima and Nagasaki.

It should be clear that metaphor is far more than "simply" a shorthand way of expressing a complex set of related ideas. It is obviously possible to list the qualities woven into a metaphorical web, as I have done. It would even be possible to attempt to untangle the threads, examining each at length and arguing for its relative weight and its relation to those around it. But while it is likely that many people would agree on the issues involved, at least in broad terms, it would remain an argument. Though I think my characterization of Wilner's metaphor is correct, it would be possible for a reasonable reader to disagree, either substantially or in terms of the details, and there is, in any event, a great deal more to be said about the matter. As anyone who has attempted an exegesis of a poem knows, there are many ways to interpret a metaphorical web. Furthermore, its meanings are not static. They shift and shimmer as one examines them. A literal gloss cannot capture this interplay, for a metaphor is, in the end, a process of thought—a dynamic balance in our minds—rather than a fixed set of meanings.

Metaphorical vs. Narrative Modes of Thought

Scientists tend to be suspicious of metaphors precisely because of this plasticity of meaning—an attitude worth mentioning because it has spread widely throughout our culture, among scientists and non-scientists alike. Science in general privileges stable meanings, the elimination of variables, and fixed mathematical relationships among those that remain. The ambiguity of metaphors, and the shifting matrix of variables they juggle, make them an inefficient medium for the production of symbolic structures suited to the manipulation of natural processes. Science builds its semiotic structures, instead, on the predictability of causal narratives (in a mixture of mathematical and "natural" languages) that unfold the same way each time.

Such repetition is not boring; in fact, scientists often apply aesthetic terms to the manifestation of phenomena predicted by the narrative structure of a pet theory, calling such underlying orderliness "beautiful." Nor is narrative repetition confined to science: the Greeks staged the same tragedies over and over; religion endlessly rehearses its peculiar narratives; Jung's archetypal narratives replay themselves from one generation to the next; and every one of us finds it comforting to refresh the "software" of consciousness by rereading a beloved book or watching a favorite movie yet again. Narrative is the foundation of causal thought, and, partly as a result of the powers granted us by the scientific method, our culture has gradually placed an increasing emphasis on narrative while decreasing its reliance on metaphor.

Metaphor has not disappeared, however; nor will it. For one thing, science is also founded, though more subtly, less comfortably, on metaphor. The webs of metaphor are the first nets scientists cast into uncharted waters, because metaphors provide a powerful way of grouping and organizing disparate phenomena. When Benjamin Franklin, for instance, proposed that electricity could be thought of as a fluid, the metaphor provided an essential conceptual framework, establishing a theoretical foundation and guiding subsequent research. But in what ways did the metaphor apply? After the first metaphorical cast, scientists compete with one another to pick out relevant threads, snip those that seem extraneous, and convert the dynamic linkages into static causal narratives. The vast majority of scientific papers are of this kind, and so the methods of science appear antithetical to metaphor, effacing its crucial role in grappling with complex, unfamiliar phenomena. Yet for all our efforts to limit, contain, and define the scope of metaphoric linkages, we cannot shake our dependence on them.

The persistence of both metaphoric and narrative modes of representation, as well as the differences between them, stems from their roots in the two primary modes of human thought. The human brain is a parallel processor—that is, it allows the "deployment of several neuron groups or several pathways to convey similar information."[18] Our neural structures allow us to process information by means of "simultaneous"[19] manipulation of redundant data sets, or different parts of a data set, or different data sets altogether. At any given instant, for example, our brains are responding to information received by our retinas, cochlea, taste buds, olfactory nerves, and tactile nerves, as well as processing abstract data regarding a book we are reading or what we want to eat for lunch, and at the same time controlling the muscles that move and focus our eyes, keep us balanced, and guide our hands, all while keeping our hearts beating, regulating our body temperature, and managing dozens of other automatic nervous functions.

Meanwhile, our conscious minds are aware of (and able to process) only a small fraction of this information. As Daniel Dennett has pointed out, "Conscious human minds are more-or-less serial virtual machines implemented—inefficiently—on the parallel hardware that evolution has provided for us."[20] Our conscious minds, therefore, experience the world sequentially, as a series of events, thoughts, and reactions, and the primary mode of this experience is narrative. We turn our daily experiences into stories around the dinner table; we employ alternate hypothetical narratives to reason through difficult decisions; we evaluate (and revise) our psychological state by examining, with the aid of a therapist, the narratives that grant us a sense of self; the very language we use is organized around sequential narrative units called sentences; even the theorems and postulates of mathematics are constructed as "If . . . then" narratives, which we combine to form elaborate causal chains echoing the process of thought that led from the simplest axiom to the most anti-intuitive conclusion.

Metaphor, on the other hand, engages the dominant parallel structures of the brain. We have already seen that metaphors form complex comparative webs, including both similarities and differences, linking vehicle and tenor. We have also seen that the links involve semiotic mapping in both directions. And we have noted that absorbing a metaphor requires rapid comparison, evaluation, and selection of relevant threads. All of these features are most efficiently and effectively handled by parallel processing. Furthermore, even if one believes it is possible to translate the simultaneous back-and-forth mapping of a metaphor into a series of narrative declarations, the power of a metaphor depends on the simultaneity of the experience. We *feel* the force of a metaphor, such as Wilner's "our thoughts, however elegant, were fire," because it allows us to hold a particular set of meanings in mind together—for the most part without additional explicit verbalization (though we may subsequently seek to express pieces of the experience in the narratives of rational thought). Metaphors are therefore tailor-made for (by) the brain's parallel processing, because such processing allows us to experience the multiple threads linking vehicle and tenor as a temporally unified mental experience.

It is worth bearing in mind that the serial functions of the brain are not synonymous with "consciousness" per se. They form, more specifically, our rational consciousness. Other animals are also conscious, but theirs is a consciousness that does not depend on narratives (or metaphors, for that matter—though it may be closer to a metaphorical way of thinking). Most animals, lacking complex language, experience each moment as a plethora of stimuli, which they sift in simultaneous relation, rather than breaking them down into components and considering each separately, sequentially. We retain this form of consciousness as well, though it is often difficult to access, being "shouted down" by the verbal facility of our serial consciousness. Meditation is one way of attempting to access the parallel portion. Metaphor is another—though because it is a function of language, it also engages the serial processes of our brain. The linking function of metaphor therefore acts not only on symbols, but on our different modes of mental processing as well. Metaphors, in effect, allow the two modes of our brains to talk to one another.

These characteristics of metaphor, along with those I outlined earlier, make them particularly suited to expressing complex relationships among phenomena, especially those involving contradictory or incompatible elements that resist representation by means of causal narratives. Nuclear weapons, as we have seen, fit this description: they combine attributes of such power and diversity that narrative representations invariably become convoluted, confusing, contradictory. Far from being undone by the bombings of Hiroshima and Nagasaki, metaphor remains an essential tool for dealing with the conceptual challenges presented by nuclear weapons, because metaphors increase the kind and range of the ways we can think about these weapons by engaging the brain's parallel functioning directly.

Metaphors, Narratives, and Nuclear Weapons

It is not surprising that metaphorical representations of nuclear weapons have proliferated in popular culture: those of us with no direct knowledge of these weapons will naturally turn to metaphor as we try to express imaginatively what we have not experienced actually. What is perhaps more surprising is that even scientists with an intimate knowledge of these weapons, despite a prejudice in favor of causal narrative, turn again and again to metaphor as they search for ways to understand and express their experiences. Alice in Wonderland; the resurrection of Christ; Columbus's arrival in the New World; the Hindu god Siva; "A red-hot elephant standing balanced on its trunk"[21]—each of these vehicles is both an expression of disorientation and an effort to reorient the self in response to a radically unfamiliar experience. Although they strive to focus on narrative renderings of their experiences, the writings of scientists who have participated in the construction of nuclear weapons offer an embarrassment of metaphorical riches that reflects the utility of the mode as a means of navigating the unfamiliar conceptual terrain.

Amidst these riches, there is one metaphor that occurs more frequently than any other, in various guises: the bomb is like god. It is an immediately compelling link, referencing the power of nuclear weapons, their apparent omnipresence, the actual omnipresence of the nuclear processes that engender them, the role of fusion in creating all of the heavier elements that make up the world, and the fact that fission and fusion reactions underlie the properties that lead us to call the sun "the giver of life." These are important associations, and they tell us a great deal about our relationship to nuclear weapons, as well as about the nature of god; but ultimately the metaphor is most compelling not in what it says, but in what it does not say.

Perhaps the most significant link between the bomb and god is the mystery that surrounds each. So far, god has eluded rational observation (although there have been many mystical sightings); nuclear weapons, being products of our own handiwork, existing as physical entities in stockpiles around the world, ought to be less resistant. We can watch film footage of nuclear explosions, look at photographs of warheads, read the accounts of *hibakusha*. Yet few of us, we pray, will ever see a nuclear weapon "incarnate"—either in use or repose. They exist for most of us only as textual entities, like god. And while fewer people may doubt the reality of nuclear weapons than question the existence of god, almost all of us who believe do so for reasons that bear an unsettling similarity to religious faith: because we grew up in households that believe, because we trust the evidence of books and films, because other people share their belief with us. Even the causal mode of scientific description that "justifies" our belief grew out of the narrative tradition of biblical exegesis.

I'm not suggesting, as I hope the sensible reader already assumes, that nuclear weapons don't exist. Nor am I suggesting that the evidence for their existence is not far more trustworthy than a religious text. But the fact remains that nuclear weapons are inaccessible to most of us. Even the military personnel and technicians who tend to these weapons have only limited access. Indeed, nuclear weapons exist within elaborate physical and institutional structures of compartmentalization and procedure designed specifically to maintain their inaccessibility—to prevent us from knowing them.

This leaves us in the curious situation of spending, worldwide, trillions of dollars to build, transport, store, and service objects we intend never to use while keeping ourselves as fully ignorant of their inner workings, external appearances, and locations as possible. Yet surely they must have, in Marx's phrase, some "use value." If so, that value would seem to be the value of a symbol—in addition to what we commonly call "symbolic value." These weapons stand for the unimaginable damage and suffering they can inflict; for the power and prestige of the society capable of building them; for the determination to "defend" a nation (or exact vengeance on its enemies) at any cost; for the Faustian bargain of scientific knowledge; for the abuse of power; for our ability to keep secrets, even from ourselves. Precisely what they mean is, of course, an ongoing debate; but there is no doubt that they do mean, and with unusual force. Indeed, it is their unrestrained capacity to mean that makes us value them so highly and fear them so intensely. While unused, they function as markers for what they might do. When used, they flood our lives with irresistible meanings—meanings so potent they destroy everything, including meaning. And so we fear them as we fear madmen, never quite confident that we understand them or what they will do. And so madmen are drawn to them as a means of imposing their mad meanings on the world.

The ability of nuclear weapons to perform these symbolic functions depends on a curious feature of the way in which they respond to metaphor. As I mentioned previously, metaphors usually involve knowledge of both the tenor and vehicle; when one has no knowledge of the tenor, the effect of a metaphor is a complete mapping of the vehicle onto the inscrutable tenor. Though we have some knowledge of nuclear weapons and their effects, it is limited and second hand, like our knowledge of god. Because of this, nuclear weapons tend (incompletely, in inverse proportion to our knowledge of them) to absorb the metaphorical vehicles applied to them, as a paintbrush absorbs paint. This renders them both difficult to know in themselves and immensely effective as symbols. They become, in effect, reservoirs of the meanings we apply to them: symbolic entities stocked with an abundance of metaphorical vehicles, ready to echo our meanings back to us, in the most profound or trivial ways: Mutually Assured Destruction, Atomic Red Hots, the Second Coming in Wrath, the Atomic Cleaners of Beaumont, Texas.

Here is the real danger inherent in metaphorical treatments of nuclear weapons. It is not that our metaphors will demean or trivialize these weapons or the experiences of those who suffered because of them. Though this is possible and should be guarded against, the more significant danger is the inverse: not the metaphors we apply to the bomb, but the ways in which we use the bomb itself as a metaphor. Like god, the bomb is a symbolic entity we have stoked with meanings that can be used to justify any argument: the U.S. is god's chosen nation; the U.S. is the great Satan; our government officials are merchants of death; our government officials are saviors, backed by their god-like bombs. Each of these dogmatic narratives can be supported by the symbolic flexibility that characterizes the inaccessible vehicle of nuclear weapons.

The problem, therefore, is not metaphor in isolation, but particular interactions of metaphor and narrative. There are good reasons to apply each mode cautiously, and the twentieth century produced plenty of examples of writers and artists who mistrusted one or the other.[22] But they did so in a wide range of contexts, and for reasons that were not peculiar to the representational challenges presented by nuclear weapons. Chief among these reasons were the insights that narrative is dangerous because it fixes meanings; metaphor because it grants them dynamic range. Yet we continue to rely on both for the essential functions they perform—functions we can't, in fact, live without. What we need to do, and urgently in the case of nuclear weapons, is find a way to use each in a measured, appropriate fashion.

The representational practices of science and religion offer convenient poles to guide our efforts. It is important, for instance, to recognize the danger of loose and imprecise applications of scientific representational modes in nonscientific contexts. While scientists do very well constructing narratives out of metaphorical insights, they are able to do so only by virtue of rigorous experimental verification of those narratives. In general, basing narratives on metaphors is imprecise, sloppy, even destructive, because it fixes certain arbitrary features of the dynamic linkages established by metaphors. On the other hand, we should also avoid metaphors lacking a solid narrative foundation. In the case of nuclear weapons, the only way to keep the bomb from being an indiscriminate and insatiable symbolic reservoir is to ground our metaphor-making in narratives from authoritative sources, including *hibakusha* and other eyewitnesses.

Religions, likewise, weave metaphor and narrative into representational webs; and their efforts to know unknowable gods offer a cautionary example for those of us who hope to apply these modes to the remote but imminent power of nuclear weapons. The importance of authorizing narratives is everywhere evident in religion: the Torah, the New Testament, the Quran, the Vedas—each offers narratives purporting to define its faith. The relative veracity of these narratives is, of course, subject to violent debate. Yet these narratives would not result in intolerance, holy fervor, and an effort to impose belief and behavior on others if they were understood to be essentially metaphoric, based on god as a symbol of a dynamic truth rather than god as a fixed, prescriptive set of narratives. Dogmatic adherents believe the narratives of their faith literally, and they are correspondingly inflexible in their attitudes and behavior. Metaphorical relations to god (favored by mystics) are less dangerous because they remain fluid, emotional rather than "logical." When a religious devotee believes faith is "reasonable" (which narrative encourages), faith turns dangerous.

Science and religion are often considered utterly incompatible disciplines, yet the metaphorical power of nuclear weapons links them inextricably, and some of the most interesting and instructive literary treatments of these weapons recognize and explore the conjunction. *A Canticle for Leibowitz* (1960), for example, begins with the discovery of a bomb shelter by Brother Francis, a novice monk, during his Lenten fast in the desert. Inside the shelter, he finds tools, a grocery list, a racing form, and a circuit diagram, among other artifacts dating to a time before a nuclear apocalypse that occurred six hundred years earlier. These artifacts belonged to Isaac Edward Leibowitz, an engineer who survived the "Deluge" and converted to Catholicism, founding a monastic order dedicated to preserving prewar knowledge from the destructive rage of the "Simpletons." Although the monks can read and write Latin, and even some English, these objects are mysteries to them. When Pope Leo asks Brother Francis if he understands the meaning of the circuit diagram, for instance, he replies: "'No, Holy Father, my ignorance is complete.' The Pope leaned toward him to whisper: 'So is ours.'"[23] The novel, therefore, depicts a world in which the symbolic meanings of nuclear weapons, unleashed in a horrific incarnation, have overwhelmed all other meanings (including, ironically, the knowledge of how the weapons were made, or even what they really were). The monks can read Leibowitz's words, but their original meanings have been erased. Everyday objects mingle indistinguishably with the specialized tools of science, all of them reduced to blank slates that the monks overwrite with their own narratives.

This metaphorization occurs in two steps. First, the people who set off the nuclear conflagration did so because they imposed certain meanings on their weapons, and those meanings, whether we consider them right or wrong from our vantage point, trumped all others. Second, the monks, encountering objects from which the nuclear bombardment had stripped all meaning, and needing to extend and strengthen their faith, metamorphosed those objects into metaphorical vehicles capable of mapping religious narratives back onto the vast experiential blank of the nuclear apocalypse. So the monks preserve Leibowitz's grocery list as a relic, using it along with all the other artifacts to make a case for their founder's canonization.

Russell Hoban's *Riddley Walker*, written twenty-one years after *Canticle*, offers a very different vision of a post-apocalyptic world, yet one in which the power of metaphor and metaphors of power play an equally important role. Like the Southeastern United States in which Brother Francis lives, Riddley's England has been utterly transformed by nuclear war. Most pre-war knowledge has been lost, and the fragments that remain are a hodge-podge, indiscriminately uniting scientific and religious narratives and metaphors. At the center of these fragments is a twentieth century informational leaflet describing a fifteenth century fresco of St. Eustace's life. The original meanings of the leaflet have been obliterated by nuclear fire, like those of the grocery list or the circuit diagram in *Canticle*, and so the descendents of the survivors (over the course of some 3,000 years) read it metaphorically, applying their own experiences, hopes, and fears to the bomb's erasures.

Out of these fragments they construct a set of narratives with which they attempt to shore up the fragments of their society. These narratives are the special property and care of an educated elite: as Riddley tells us, "You wunt have seen the woal thing wrote out without you ben a Eusa show man or connexion man or in the Mincery. No 1 else is allowit to have it wrote down the same which that dont make no odds becaws no 1 else knows how to read."[24] The show men travel from community to community, retelling their stories by means of elaborate Punch and Judy shows, which are then interpreted by the local connexion man. The cycle of retelling and reinterpretation binds the various communities in a web of shared meanings, creating a larger political entity in much the same way individual religious communities are bound into a single religious institution by the semiotic glue of sanctioned metaphors and narratives.

The "Eusa story" in *Riddley Walker* is neither quite scientific nor fully religious, but it combines clear elements of both. It includes, for instance, the pre-apocalyptic discipline of chemistry, now known as "chemistery"—part scientific discipline, part religious mystery—that tells, among other things, of "the Littl Shynin Man the Addom." Here and elsewhere, Hoban brilliantly reconstructs the historical links between religious myth and scientific theory, in this case by merging pieces of the narrative of Adam and Eve with bits of atomic theory to create a kind of doubled metaphor. By "doubled" I mean that it functions as a metaphor both within the world of the novel—a narrative figure that serves as a vehicle for the mysterious forces of the "1 Big 1"—and for the reader—a trope linking the religious and scientific narratives of Adam and atom. In each case, the metaphor is rich with meaning and practical implications. For Riddley's people, it represents both a means of social cohesion and a step on the long road of (re)constructing (religious, scientific) narratives that constitute understanding of the natural world. For us, it reveals some of the ways metaphors and narratives function—as modes of finding and making meaning in the world; as means of imagining how our world might be laid waste by the ways we have construed it; and as tools for reimagining a world in which the threat of nuclear weapons is not simply a matter of megatons but of how we think and write and speak.

A crucial component in the doubling of Hoban's metaphor is that we know things that Riddley and his compatriots do not. The original meanings of St. Eustace, Adam, and atom are still accessible to us, while Riddley's knowledge is limited to the reconfigured narratives and metaphors of the novel. We therefore see two sets of meanings "side by side" (though separated by 90 pages): the originals and those revised by nuclear war. From the leaflet:

> 1. At the bottom of the painting St Eustace is seen on his knees before his quarry, a stag, between whose antlers appears, on a cross of radiant light, the figure of the crucified Saviour.[25]

And from the Eusa story:

> 8. In the dark wud Eusa seen a trak uv lyt he follert it. He cum tu the Hart uv the Wud it wuz the Stag uv the Wud it wuz the 12 Poynt Stag stud tu fays him & stampin its feat. On the stags hed stud the Littl Shynin Man the Addom in be twean thay horns with arms owt strecht & each han holdin tu a horn.[26]

In a fascinating passage, unfortunately too long to reproduce here, a character named Goodparley offers an interpretation of this doubled text that reveals both how far he and Riddley are from grasping its pre-apocalyptic meanings and yet how slowly and surely they are constructing new, viable metaphoric and narrative structures that will lead from an unlikely foundation in the life of St. Eustace to the rediscovery of gunpowder. Our ability to see both the old and the new structures of meaning allows us to recognize not only what we stand to lose, but the narrative and metaphoric processes by which it came to be created in the first place.

Furthermore, and most disturbingly, both *Canticle* and *Riddley* are ultimately mirrors in which what we gaze at is ourselves. Brother Francis and Riddley Walker are fictional; what meaning we find in them is our own. After our initial impulse of smug superiority in response to the obvious ignorance of these post-apocalyptic simpletons, an attentive reader will begin to recognize parallels deeper and more compelling than the surface differences. Do we really know more about nuclear weapons than the people who inhabit a world utterly reshaped by their desctructive power? Their ignorance is an extension—consequence and symbol—of ours. Atom and Addom are equally powerful, equally inadequate signifiers of what remain for most of us matters of "chemistery" better left to others to decipher. And in our acceptance of that ignorance we unknowingly embrace a more lasting ignorance: that of Francis and Riddley; the silence of fallout; the potential erasure of all we think we know.

Conclusion

All of this suggests a number of answers to the question I asked earlier: what can our metaphoric representations add to the narratives of those with firsthand knowledge of nuclear weapons? Without Hoban's novel (and other metaphoric explorations of all sorts), we would be more likely to accept the metaphors of eyewitnesses as exclusively authoritative, to treat narratives based on these metaphors as fixed, to find ourselves stuck in outmoded ways of thinking. Einstein famously said, "The unleashed power of the atom has changed everything save our modes of thinking, and we thus drift toward unparalleled catastrophe."[27] By "modes" he did not mean either metaphors or narratives in themselves, but rather the particular ways we used them immediately following World War II. Einstein saw that we were allowing ourselves to become dependent on old, obvious metaphors and the overly rigid narratives they inspired. But *Riddley Walker*, *A Canticle for Leibowitz*, and works like them grant us a second perspective—imaginary, yes, but based on a wide-ranging knowledge of the facts as we know them. In doing so, the novels function as extended metaphors themselves, linking the narratives and metaphors of our experience with those of Hoban's and Miller's imagined worlds. Such metaphoric acts of imagination can help us to see our own "modes of thinking" more clearly, as well as where they might lead, and ultimately contribute to a potential transformation, or at least revision, of how we think and act in response to nuclear weapons.

Notes

[1] Christopher Felver, "Donald Judd's Marfa, Texas," 1998.

[2] Adorno wrote, "Perennial suffering has as much right to express itself as the martyr has to scream; this is why it may have been wrong to say that poetry could not be written after Auschwitz." *Negative Dialectics*, trans. by Dennis Redmond, http://www.efn.org/~dredmond/ndtrans.html, 2002.

[3] These are, of course, enormous questions. In the hope of finding a way toward practical insight, I will limit the scope of the issues I consider, focusing on the functions of metaphor relative to nuclear weapons and leaving issues of representing and responding to the Holocaust to others. I will also look at metaphor in its verbal manifestations rather than its use in the plastic arts.

[4] Nuclear Weapon Archive, "The B-41 (Mk-41) Bomb: High yield strategic thermonuclear bomb," last updated 10/21/97, http://nuclearweaponsarchive.org/Usa/Weapons/B41.html.

[5] Because these weapons had never before been used in combat, the U.S. military was eager to gather data, so each aircraft was accompanied by two observation planes; these planes were not essential to the mission. In addition, one scout plane flew ahead to each of the three potential targets to check visibility.

[6] United States Navy, "Fact File: Fleet Ballistic Missile Submarines," http://www.navy.mil/navydata/fact_display.asp?cid=4100&tid=200&ct=4; Lockheed Martin, "Navy Team Celebrates 10th Anniversary of Trident II D5 Missile System," http://www.lockheedmartin.com/wms/findPage.do?dsp=fec&ci==12491&rsbci=0&fti=0&ti=0&sc=400; Nuclear Weapon Archive, "The W88 Warhead: Intermediate yield strategic SLBM MIRV warhead," http://www.nuclearweaponarchive.org/Usa/Weapons/W88.html.

[7] United States Navy, "Fact File: Trident Fleet Ballistic Missile," http://www.navy.mil/navydata/fact_display.asp?cid=2200&tid=1400&ct=2.

[8] When we speak of "ease," it is worth remembering that 120,000 people worked for two years to build the first three atomic bombs and that an ongoing massive collaboration between government, military, academic, and industrial institutions is required to design, produce, and maintain these weapons.

[9] Bulletin of the Atomic Scientists, "5 Minutes to Midnight: Nuclear," http://www.thebulletin.org/minutes-to-midnight/nuclear.html

[10] Nuclear Age Peace Foundation, "Nuclear Stockpiles," http://www.nuclearfiles.org/menu/key-issues/nuclear-weapons/basics/nuclear-stockpiles.htm

[11] Sight is a process involving the transfer of information from an object to an eye by means of huge numbers photons. Atoms are over a thousand times smaller than the smallest wavelength of visible light, so an interaction between a photon and a subatomic particle can convey only partial information—and none that our brains can register. The photon's impulse changes the behavior of the particle it is "reporting on" anyway.

[12] Despite the scientific emphasis on literal over metaphorical language, it is precisely this lack of analogs that proves most troubling to scientists, since the process of constructing literal descriptions of the world involves initial forays into highly metaphorical conceptions. I will return to this point later.

[13] Radiation therapy was first used to treat cancer in 1899. American Cancer Society, "The History of Cancer," revised 3/25/02, http://www.cancer.org/docroot/CRI/content/CRI_2_6x_the_history_of_cancer_72.asp?sitearea=CRI.

[14] Michihiko Hachiya, *Hiroshima Diary* (Chapel Hill, NC: University of North Carolina Press, 1955), 43.

[15] I'm coining "memetic" based on Richard Dawkins notion of "memes" as patterns of information (ideas, metaphors, behaviors) that reproduce throughout a culture. We are the medium of their reproduction, as of our genes'. And as our genes are, collectively, a blueprint of our bodies, memes structure our non-inherited behaviors and identities. See Richard Dawkins, *The Selfish Gene*, 3rd ed. (Oxford: Oxford University Press, 2006).

[16] This metaphor is, itself, an excellent example of the complex, ongoing processing involved in the creation and consumption of symbolic linkages, having been the subject of considerable exploration and debate since it was formulated so memorably by William James in his *Principles of Psychology* (1890).
[17] Eleanor Wilner, *Sarah's Choice* (Chicago: University of Chicago Press, 1989).
[18] Eric R. Kandell, James H. Schwartz, Thomas M. Jessell, *Principles of Neural Science*, 4th edition (New York: McGraw Hill, 2000), 34.
[19] The term is vexed, both relativistically and when applied to neural processing, in the latter of which there is approximately a 50 msec range during which two events are effectively "simultaneous." See Daniel C. Dennett and Marcel Kinsbourne, "Time and the Observer: The Where and When of Consciousness in the Brain," in *The Nature of Consciousness*, eds. Ned Block, Owen Flanagan, and Güven Güzeldere (Cambridge, MA: MIT University Press, 1997), 143.
[20] Daniel Dennett, *Consciousness Explained* (Boston: Little, Brown and Company, 1991), 218.
[21] Robert Serber, correspondence with the author, March 8, 1993; Victor Weisskopf, *The Joy of Insight: Passions of a Physicist* (New York: Basic Books, 1991), 152; Arthur Compton, *The Cosmos of Arthur Holly Compton*, ed. Marjorie Johnston (New York: Alfred A. Knopf, 1967), 248; Robert Oppenheimer, quoted in Richard Rhodes, *The Making of the Atomic Bomb* (New York: Simon and Schuster, 1986), 676; Otto Frisch, *What Little I Remember* (Cambridge: Cambridge University Press, 1980), 164.
[22] James Joyce, T.S. Eliot, and Samuel Beckett come to mind as examples of writers who questioned conventional narratives; Ernest Hemingway, Theodor Adorno, and Donald Judd mistrusted metaphor.
[23] Walter M. Miller, Jr., *A Canticle for Leibowitz* (Philadelphia: J.B. Lippincott, 1973), 112.
[24] Russell Hoban, *Riddley Walker* (New York: Summit Books, 1980), 29.
[25] Ibid., 123.
[26] Ibid., 31.
[27] Albert Einstein, *The New York Times* (25 May 1946), p. 13, col. 5.

Contributors

Michael Blouin is an Assistant Professor in English and the Humanities at Milligan College. His research interests include atomic culture, literary theory and criticism, and transnational studies. He is the author of a forthcoming book entitled *Japan and the Cosmopolitan Gothic: Specters of Modernity* (Palgrave Macmillan, 2013).

John Canaday approaches nuclear studies from both analytic and creative perspectives. He is the author of *The Nuclear Muse: Literature, Physics, and the First Atomic Bombs* as well as a series of poems in the voices of the men and women—including scientists, spouses, laborers, locals, and military personnel—involved in the Manhattan Project (a sample of which can be found at http://atlengthmag.com/poetry/from-critical-assembly/).

Daniel Cordle is Reader in English and American literature at Nottingham Trent University. He has worked extensively on nuclear culture and is the author of *States of Suspense: The Nuclear Age, Postmodernism and United States Fiction and Prose* (Manchester University Press, 2008). He also has a broader interest in cultures of science and technology. Among his other publications is *Postmodern Postures: Literature, Science and the Two Cultures Debate* (Ashgate, 1999).

Joseph Dewey is an Associate Professor of Contemporary American Literature for the University of Pittsburgh System. He is the author of four studies of contemporary American literature that have looked at the relationship between literature and religion: *In a Dark Time: The Apocalyptic Temper of the American Novel in the Nuclear Age* (1990); *Novels from Reagan's America: A New Realism* (1997); *Understanding Richard Powers* (2001); and *Beyond Grief and Nothing: A Reading of Don DeLillo* (2006). His essays, more than 100, have appeared in a variety of literary journals over the last thirty years. His subjects of interests have included Edward P. Jones, Rick Moody, Marilynne Robinson, Harry Crews, Jeffrey Eugenides, and Frederick Buechner. Dr. Dewey holds a B.A. in Literature from Villanova and an M.A. and Ph.D. in Modern American Literature from Purdue. He is completing a manuscript on Michael Chabon for the South Carolina University Press.

Bradley J. Fest received his MFA in poetry from the University of Pittsburgh, where he is now a Visiting Instructor and PhD candidate studying nineteenth through twenty-first century American literature. He is currently completing his dissertation, "The Apocalypse Archive: American Literature and the Nuclear Bomb." His work has appeared in *boundary 2* and *Studies in the Novel*. His poetry has been published in *Spork, Open Thread, BathHouse, Flywheel*, and elsewhere. He blogs at *The Hyperarchival Parallax*.

Jessica Hurley is a doctoral candidate and Benjamin Franklin Fellow at the University of Pennsylvania. She holds a BA (Hons) from the University of Oxford and an MA from the University of Sussex. Her research focuses on the affective life of apocalyptic rhetoric in contemporary American literature and culture.

William M. Knoblauch is currently an Assistant Professor of History at Finlandia University (Hancock, MI). Previously, he was a fellow of the Contemporary History Institute (Athens, Oh), and received his Ph.D. in American History from Ohio University in 2012.

Mark Pedretti holds a Ph.D. in Rhetoric from the University of California, Berkeley, and he is currently a Lecturer in the Department of English at Case Western Reserve University. He is currently at work on a book about British and American late modernism in relation to the early atomic age.

Aaron Rosenberg is a doctoral candidate in English at Cornell University, where he specializes in 20th century literature. His work has appeared in the *Journal of Modern Literature*.

Paul K. Saint-Amour is Associate Professor of English at the University of Pennsylvania and has been a fellow at the Stanford Humanities Center, the Society for the Humanities at Cornell, and the National Humanities Center. His book *The Copywrights: Intellectual Property and the Literary Imagination* (Cornell, 2003) won the MLA Prize for a First Book. Saint-Amour edited *Modernism and Copyright* (Oxford, 2011) and is co-editor, with Jessica Berman, of the Modernist Latitudes series at Columbia. He is currently at work on a book entitled *Archive, Bomb, Civilian: Modernism in the Shadow of Total War*.

Patrick B. Sharp is Professor and Chair of the Department of Liberal Studies at California State University, Los Angeles. He has published Nuclear Criticism focused on Darwinism, race, and gender in journals such as *Twentieth-Century Literature* and *Science Fiction Film and Television*. He co-edited the anthologies *Darwin in Atlantic Cultures* (Routledge 2009) and *Practicing Science Fiction* (McFarland 2010), and his 2007 monograph *Savage Perils: Racial Frontiers and Nuclear Apocalypse in American Culture* has just been reissued in paperback by the University of Oklahoma Press.

Morgan Shipley is a PhD Candidate in the Program in American Studies and an Instructor in the Center for Integrative Studies in the Arts & Humanities at Michigan State University. He received his BA from DePaul University and an MA from the University of Chicago. His dissertation investigates the relationship between psychedelic consciousness, mystical frames of interpretation and a spiritual ontology of altruism as envisioned in the work of Aldous Huxley, Stephen Gaskin, Allen Ginsberg, Timothy Leary and Alan Watts. He has articles forthcoming in *Utopian Studies* and, along with Michael Blouin, in *The Journal of Popular Culture*.

Jeff Smith is the author of *Unthinking the Unthinkable: Nuclear Weapons and Western Culture* and *The Presidents We Imagine: Two Centuries of White House Fictions on the Page, on the Stage, Onscreen, and Online*. A news reporter and commentator, television news consultant and theater director, he has taught at the University of Illinois, UCLA and the University of Southern California, holds a Ph.D. in English and American Studies from the University of Chicago and an MFA in Theater, Film and Television from UCLA, and has studied and taught the Cold War and cultural history as a Fulbright Fellow in Eastern Europe and the UK and and a Visiting Fellow at Oxford University's Rothermere American Institute.

Jack Taylor is PhD candidate in American Studies at Michigan State University. His interests include the intersection of African American literature and philosophy with an emphasis on death and racial violence, African American photography and visual culture, political theory with an emphasis on necro and biopolitics, and American Anthropology. He is currently working on a dissertation concerning representations of death in African literature and art.

Julie Montana Williams is a PhD candidate and Graduate Student Instructor in the English department at the University of New Mexico. Her work focuses on 19th and 20th Century Western American Literature, Literature and the Environment, Feminist Theories, and the Atomic Age. She has recently published on landscape and Southwestern cultures in Willa Cather's *Death Comes for the Archbishop in Plaza: Dialogues in Language and Literature*. She was awarded the 2011-2012 Hector Torres Memorial Fellowship through the University of New Mexico's Center for Regional Studies to pursue research on the history of the Los Alamos community, its military and scientific institutions, and the role that atomic production has played in the culture of New Mexico and the greater Southwest.

INDEX

9/11 (September 11, 2001 Terrorist Attacks) 26, 53, 58, 124, 133, 135, 136, 137, 159, 211, 226, 233
A Canticle for Leibowitz 68, 69, 70, 72, 73, 74, 77, 79, 85, 90, 91, 94, 100, 108, 131, 266, 269
Allegory 2, 116, 147, 172, 184, 204
Anachrony 180, 182, 184
Anathem 88, 90, 91, 92, 94
Anderson, Benedict 19, 36
Apocalypse 10, 11, 60, 63, 73, 85, 88, 109, 120, 122, 123, 124, 129, 131, 132, 133, 134, 135, 144, 182, 193, 205, 207, 211, 213, 215, 216, 221, 223, 224, 226, 266
Archive 4, 5, 12, 52, 55, 65, 66, 67, 73, 74, 75, 76, 77, 83, 84, 85, 86, 87, 88, 90, 91, 92, 93, 94, 95, 96, 97, 98, 100, 101, 102, 202
Blanchot, Maurice 14, 49, 50, 51, 53
Buell, Lawrence 232, 234, 235, 236, 242, 243, 246
Burntime 131, 132, 141
Bush, George W. 8, 9, 31, 42, 58, 112, 137, 138
Cancer 152, 154, 155, 156, 157, 232, 237, 238, 245, 246, 253, 255
Carson, Rachel 213, 214, 215, 221, 226
Catching Fire 211, 221
Catholic 27, 68, 71, 72, 80
Catholicism 74, 266

Chase, David 106, 107, 120
Chow, Rey 5
Clausewitz, Carl von 196
Cold War 3, 4, 5, 6, 7, 8, 9, 11, 12, 16, 18, 25, 26, 27, 28, 29, 31, 32, 35, 36, 39, 48, 52, 54, 56, 60, 67, 68, 82, 83, 85, 87, 89, 94, 95, 96, 97, 99, 106, 108, 111, 112, 113, 122, 124, 125, 126, 127, 129, 130, 131, 132, 134, 138, 143, 144, 145, 152, 153, 155, 158, 159, 162, 168, 169, 185, 186, 187, 191, 195, 202, 211, 213, 214, 215, 216, 218, 219, 222, 226, 231, 232-39, 241, 242-44, 247, 253
Collins, Suzanne 211, 213, 216, 217-226
Containment 115, 174, 231, 246, 247
Contamination 158, 214, 215, 240, 245, 253
Cornell Conference 2, 7, 11, 13, 45, 48, 100, 167, 192
Coviello, Peter 63, 78
Cuban Missile Crisis 54, 125, 129, 166, 214
Cyclical 68, 76, 194, 202
Darwin, Charles 218, 219, 220, 221, 225
Darwinism 12, 211, 218
De Kerckhove, Derrick 193, 195
Death Drive 12, 53, 62, 63, 74, 75, 76
Deconstruction 3, 4, 26, 45, 96, 202
DeLillo, Don 104, 114, 168, 194

Derrida, Jacques 2, 3, 5, 6, 7, 8, 10, 11, 45, 46, 47, 49, 50, 51, 52, 53, 55, 56, 57, 63, 64, 65, 66, 67, 74, 75, 77, 79, 83, 84, 86, 87, 95, 96, 97, 100, 101, 143, 145, 146, 163, 193, 194, 202, 204, 205, 206, 207
diacritics 45, 53, 167, 187, 192, 193
Downwinders 152, 153
Dr. Strangelove; Or, How I Learned to Stop Worrying and Love the Bomb 26, 28, 56, 89, 122, 128, 255
Ecological 82, 86, 90, 97, 98, 103, 211, 213, 215, 216, 218, 221, 223, 226, 236
Einstein, Albert 269
Elision 180
Eschatology 94, 193
Evolution 213, 215, 218, 219, 224, 225
Fallout 9-10, 13, 48, 54, 118, 125, 132, 153, 155, 157, 213, 214, 234, 253, 269
Fallout 3 125, 134, 135, 136, 137
First nuclear age 4, 83-88, 96
First-person Shooter 130, 134, 135, 137, 140
Frontier 91, 131, 141, 205, 211, 216, 217, 218, 219, 221, 222, 223, 224, 225, 226
Futurism 60, 61, 62, 64, 65, 67, 68, 70, 76, 77, 78
Gaming 131, 134, 135, 136
Gender 22, 43, 211, 218, 219, 221, 225, 226
Hawkes, John 172, 180, 181, 182, 184
Hersey, John 1, 13, 172, 173, 174, 175, 176, 178, 179, 180, 181, 182, 183, 184, 189
Hibakusha 149, 150, 151, 152, 163, 164, 179, 180, 245, 252, 263, 265

Hiroshima 1, 172, 173, 174, 175, 176, 177, 178, 179, 180, 251
Hiroshima Bugi: Atomu 57 144, 146, 147, 148, 149, 152, 153, 156, 161
Hoban, Russell 100, 255, 266, 267, 268, 269
Hyperarchival 94, 96, 97, 102
Infinite Jest 192, 194-206
Instrumentalism 16, 18, 36
Instrumentalization 176
Internet 12, 81, 82, 83, 89
Iraq 28, 30, 35, 112, 211
Jameson, Fredric 82, 97, 168, 169, 171, 177, 187, 203
Jaspers, Karl 49
Kahn, Herman 26, 27, 29, 31
Kenzaburo, Oe 149
Klein, Richard 47, 55, 170, 187, 192
Leakage 6, 231, 232, 246
Legacy Waste 230, 247
Lippit, Akira 6
Madsen, Michael 230, 231, 246
Mamet, David 88, 89, 90, 94, 96, 101
Manhattan Project x, 125, 166, 195, 232, 245
Masco, Joseph 232, 234, 242
Mason, Bobbi Ann 232, 247
McCarthy, Cormac 6, 60, 91, 116-120, 133, 144, 158, 159
Metaphor xiv, 50, 81, 109, 149, 154, 155, 234, 239, 250, 255, 256-269, 271
 Metaphorical 251, 255-257, 259-262, 264-266, 270
 Metaphorization 266
Miller Jr., Walter M. 68, 69, 70, 74, 75, 85, 108, 131, 269
Millet, Lydia 232, 241
Mockingjay 211, 221, 222
Modernism 167, 168, 169, 170, 176, 177, 180, 185, 187
Mutual Assured Destruction (M.A.D.) xi, xiii, 81, 82, 86,

87, 88, 109, 162, 194, 198, 211, 221, 222, 226, 264
Nadel, Alan 167, 174, 175, 176, 177
Nature Writing 152, 232, 235, 236, 237
Nineteen Eighty-Four 23, 31, 41
"No Apocalypse, Not Now" 3, 46, 63, 64, 65, 67, 75, 83, 143, 163, 202, 205
NORAD 81, 82, 102, 122, 123, 138, 195
Norris, Christopher 48
Nostalgia 16, 32, 33, 34, 38, 40, 53, 108, 110, 129, 134, 135, 138, 233
Nuclear Culture 230, 231, 232, 233, 241, 243, 244, 247
Nuclear Modernism 170, 171, 172, 184, 185, 186
Nuclear Referent 4, 60, 63, 82, 83, 85, 86, 90, 96, 101, 144, 145, 146, 153, 155, 159, 163
Nuclear Technology 11, 167, 254
"Nuclear Uncanny" 12, 54, 232, 241, 242, 244
Nuclear War xiii, 3, 25, 29, 45-51, 54, 55, 56, 64-66, 68, 71, 73, 81, 82-88, 91, 94, 97-98, 122-132, 134, 138, 143-146, 153, 155, 159, 166, 192, 194, 198, 217, 231, 233, 236, 242, 266, 268
Nuclear War 123, 127, 128, 131
Nuclear Weapons x, xi, xiv, xvi, 6, 10, 11, 16, 18, 27, 46, 48, 51, 54, 61, 63, 76, 77, 78, 123-125, 133, 137, 138, 143-146, 149-150, 152, 153, 155, 159, 161-163, 166, 167, 171, 179, 194, 195, 211, 213, 215, 216, 222, 223, 224, 226, 244, 251-255, 262-266, 268-269
"Nuclear Winter" 117, 124, 130, 132, 134, 135, 159
Nuclearism 167

Nuclearity 234, 236, 238, 240, 241, 243, 244, 245, 246, 247
"Palimpsest" 88, 92, 93, 94
Onkalo 230, 231, 246
Oppenheimer, Robert x, xi, xv, xvi, xvii, 232, 233, 241, 243
Orwell, George 31, 41
Osmosis 233, 234
Peace 5, 22, 56, 146, 148, 149, 150, 151, 154, 155, 241
Periodization 167, 168, 169, 170, 174, 185, 186
Postmodern 82, 86, 90, 146, 167, 168, 169, 170, 171, 174, 175, 176, 177, 185, 186, 194, 198
Postmodernism 167, 168, 169, 170, 174, 175, 176, 185
Post-nuclear 32, 86, 87, 88, 100, 124, 132, 168, 174, 181, 202
Queer Temporalities 60, 61, 62, 63, 67, 75
Queer theory 5, 62, 63, 66, 78
Radioactive 125, 129, 131, 132, 133, 157, 172, 201, 214, 230, 253, 255
Reagan, Ronald 2, 17, 25, 27, 29, 58, 109, 123, 124, 126, 127, 129
Refuge 144, 152-155, 157, 158, 161, 232, 233, 235, 236, 237, 240, 241, 244, 245, 246
Remainderless 48, 67, 83, 84, 86, 87, 91, 95, 96
Reproductivism 62, 67
Ruthven, K.K. 3, 4, 12, 57, 109, 110
Schell, Jonathan 8, 9, 16, 40, 59, 60, 62, 64, 67, 99, 123, 192, 206
Schwenger, Peter 4, 5, 100
Science Fiction 211, 212, 213, 219, 226
Science xiii, xv, xvi, 16, 64, 90, 96, 109, 125, 130, 131, 132, 199, 211, 213, 253, 255, 256, 259, 260, 265, 266

Scientific xi, xiii, xiv, 6, 7, 11, 45, 64, 67, 68, 74, 108, 130, 135, 176, 213, 260, 263, 264, 265, 267, 270
Second Nuclear Age 83, 84, 88, 89, 95, 96, 103
Shakespeare, William 19, 20, 21, 22, 24, 25, 29, 32, 34, 40
Stephenson, Neal 88, 90, 92, 94, 96
Strategic Defense Initiative (SDI) 25, 126, 129, 132, 138
Stross, Charles 88, 92, 94, 96
"Survivance" 52, 65, 67, 145, 147, 148, 149, 151, 153
Symbol 38, 68, 96, 97, 205, 206, 242, 245, 247, 263, 265, 268
Symbolic xii, xiv, 5, 51, 55, 64, 65, 68, 69, 70, 82, 105, 107, 109, 117, 259, 263, 264, 265, 266, 271
Techno-scientific 47, 74
Temporal 10, 48, 55, 75, 91-94, 170, 171, 177-180, 182, 184, 187, 188, 189, 201, 203, 205, 246
Temporality 12, 49, 61-62, 69, 72, 80, 87, 88, 92, 172, 177, 179-184, 187, 193, 197, 201, 202, 203
Terror 9, 18, 22, 24, 28, 29, 49, 50, 53, 82, 112, 137, 179, 200, 206, 232, 244
Terrorism 9, 83, 113, 159, 215
Terrorist 8, 11, 28, 37, 111, 114, 124, 125, 133, 194, 197, 208, 215, 226, 233

Textuality 51, 88, 101, 170, 174, 175
The Fate of the Earth 16, 59, 64, 192
The Hunger Games 10, 211, 213, 215, 216, 218, 226
The Road 6, 12, 60, 91, 116, 117, 120, 134, 144, 158-159, 161, 162
The Sopranos 6, 104, 106
Toxic Discourse 232, 235, 236, 240, 241, 242, 243
Toxicity 10, 213, 223, 226, 230, 236, 239, 241-243
Turner, Frederick Jackson 216, 217, 218, 219
Virilio, Paul 192, 198-200, 205
Vizenor, Gerald 144, 146, 147, 148, 152, 153, 156, 158
Riddley Walker 100, 255, 266, 267, 268, 269
Wallace, David Foster 194, 195, 202, 203, 204, 207, 209
Wallace, Molly 53, 103
War on Terror 29, 53, 82, 137, 200, 232, 244
Wargames 122, 123, 124, 126, 130, 133, 138, 192
Williams, Terry Tempest 144, 152, 153, 154, 155, 156, 157, 158, 232, 235, 236, 237-241, 245, 246
Wilson: A Consideration of the Sources 88, 89, 90, 92, 94, 102